Industrial Management

– tools and techniques

MATS ENGWALL

ANNA JERBRANT

BO KARLSON

FREDRIK LAGERGREN

PER STORM

PAUL WESTIN

Translation: Marcia Halvorsen

Studentlitteratur

Original title: *Industriell ekonomi – metoder och verktyg*
© Studentlitteratur, Lund 2014

Art. No 38967
ISBN 978-91-44-10788-2
First edition
1:1

© The authors and Studentlitteratur 2016
studentlitteratur.se
Studentlitteratur AB, Lund

Translation: Marcia Halvorsen
Cover design: Jens Martin/Signalera
Cover illustration: Shutterstock/vs148

Printed by Interak, Poland 2016

We dedicate this book to our friend and colleague

Albert Danielsson, 1930–2014

Professor in Industrial Economics and Management at KTH Royal Institute
of Technology, Stockholm, Sweden, 1969–1996
Founder of the MSc Program in Industrial Engineering and Management
at KTH Royal Institute of Technology, Stockholm, Sweden

CONTENTS

FOREWORD TO THE ENGLISH EDITION

This book was written by a group of teachers and researchers, present and former, at the Department for Industrial Economics and Management (Indek) at KTH Royal Institute of Technology, Stockholm, Sweden. As we are engineers, we have written a book on Industrial Management for engineers and engineering students. Therefore, the book deals with a great many management, business, and financial concepts as well as the tools and techniques engineers need to understand so that they can plan, develop, and control industrial operations and businesses.

The book is primarily aimed for use in basic courses in Industrial Management for engineering students in Sweden and the other Nordic countries. Thus, without claiming to provide the complete picture, the book presents the most fundamental concepts, tools, and techniques in the Swedish tradition of Industrial Management. However, despite local differences compared with other national traditions or juridical systems, the basic models and principles are generally the same in various parts of the world. The Swedish business terminology was translated into English using *FARs engelska ordbok* (14th ed., FAR Akademi 2011), which is the official English dictionary of the Swedish Institute of Authorized Public Accountants (FAR).

The book is a development and update of the book *Industriell ekonomi* that (partly) the same group of authors published in 1998. Writing this book was very much a collective act. In the 1998 book, Professor Albert Danielsson (Head of the Department for Industrial Economics and Management at KTH Royal Institute of Technology from 1969 to 1996) was the main influence on the book that reflects his approach to management and business studies. In the years since, practices in the area of Industrial Management have continued to develop. Similarly, our understanding of the area has also increased.

Writing a book such as this requires a great deal of support. We especially thank our author-colleagues from the first book project: Magnus Aniander, Henrik Blomgren, Fredrik Gessler, and Jacob Gramenius. Without their help, this new book would not have been possible. We also send our thanks to Eric Rehn, at our publisher, for his excellent assistance. His enormous patience made this book possible.

Stockholm, May 2016

Mats Engwall, Professor   *Anna Jerbrant,* PhD
Bo Karlson, PhD   *Fredrik Lagergren*, PhD
Per Storm, PhD, Docent   *Paul Westin*, MSc

ABOUT THE AUTHORS

Mats Engwall is a Professor of Industrial Management at KTH Royal Institute of Technology in Stockholm. He has a Master of Science in Aerospace Engineering and a PhD in Industrial Engineering and Management from KTH. He is an Associate Professor of Business Administration at the Stockholm School of Economics. His research interests include organizational activities, management of industrial research and development, servicification, service companies, business models, and value creation. He is a member of the Royal Swedish Academy of Engineering Sciences, the Swedish Project Academy, and the European Academy for Industrial Management. He is chairman of the Scandinavian Academy of Industrial Engineering and Management.

Anna Jerbrant is an Associate Professor and Head of Division at the Department of Industrial Economics and Management at KTH Royal Institute of Technology in Stockholm. She has a PhD in Industrial Engineering and Management from KTH. Her research focus is the management of project-based organizations. Since 2010 she has been the Program Director of KTH's Master of Science in Industrial Engineering and Management. In this position she is responsible for the daily operations of the program as well as its strategic development. Between 2013 and 2015 she was one of five education developers at the School for Industrial Engineering and Management at KTH.

Bo Karlson is a Senior Lecturer and Studies Director at the Department of Industrial Economics and Management at KTH Royal Institute of Technology in Stockholm. He has a PhD in Industrial Engineering and Management from KTH. In addition to his many years of teaching and researching in the area of industrial management, he has had many roles at KTH such as

Business Developer and Project Manager for KTH Professional Education and Manager of KTH's Business Liaison. He has also been Manager of the research centre Wireless@KTH. For a few years he worked with a small start-up company in the mobile phone sector.

Fredrik Lagergren has a PhD in Industrial Engineering and Management from KTH Royal Institute of Technology in Stockholm. He is the CEO of Samarbetande Konsulter AB (SAM). From 2004 to 2016 he was the Program Director at KTH's Executive School where he developed and implemented strategic development programs for senior managers working in sectors such as utilities, public transport, telecommunications, banking, public management/leadership, etc. From 2006 to 2015 he was the Chief Executive Officer of KTH's Executive School. He is a member of the Royal Swedish Academy of Engineering Sciences.

Per Storm is the Chief Executive Officer of the publicly owned exploration firm, Copperstone Resources AB, and an Associate Professor in the Innovation Processes of the Mining and Steel Industries at KTH Royal Institute of Technology. He is the Chairman of EccaNordic AB and a Board Member of the research organisation Institute for Security and Policy Development (ISPD). He was previously Managing Director at the mining strategy and policy company Raw Materials Group RMG AB. He was also a Board Member of two exploration and mining development companies: Norrliden Mining AB and Nordic Iron Ore AB. He has a PhD in Industrial Engineering and Management and a PhD in Metals Production Technology, both from KTH. He is currently a member of the Royal Swedish Academy of Engineering Sciences.

Paul Westin has a Master of Science in Industrial Engineering and Management from KTH Royal Institute of Technology in Stockholm. Currently he is the Counselor for Energy at the Swedish Embassy in Jakarta, Indonesia. Previously he was the Deputy Director at the Swedish Energy Agency where he was responsible for scientific studies on energy. He was also on the Board of Trustees at the Oxford Institute for Energy Studies as well as on the Board of World Energy Council Sweden and the Board of Norrenergi AB.

CHAPTER 1

Introduction

Understanding the interrelationship between technology and business is of significant importance for all actors engaged in technologically intensive operations. Identifying business opportunities, making investment decisions, estimating production costs, designing work systems, organizing project teams, exercising cost control, implementing production changes, and conducting innovative technology development projects are typical examples of business activities that engage engineers. Consequently, management for engineers, the field we call *Industrial Management* (in Swedish: *Industriell ekonomi*), is an old and well-established academic discipline. The label, or name, of the discipline may vary among countries and universities, but education and research on management and business related to technology and engineering are found at almost all major European universities of technology. In Sweden, Industrial Management was established as an academic field in 1912 when the first professor of the field was employed at KTH Royal Institute of Technology in Stockholm.

Knowledge of the interactions between business/management and technology/engineering is in great demand in industry. Engineers must have the skills to understand the technical implications of management and financial decisions as well as the business and managerial implications of various technical decisions. This requires the capability for understanding and cooperating with technical specialists as well as business administrators, accountants, and financial experts. Most of today's employers expect their engineers to manage and support their work using financial terminology, to read and write financial reports, to estimate costs using established accounting techniques, etc. Consequently, this book is written for engineers and engineering students who will face the many financial and managerial challenges of the modern business world.

1.1 Industrial Management as an academic field

Industrial Management revolves around the design, development, and implementation of effective and efficient value-creating processes. In general, management as a subject of study is an *aspect* of society, much like other disciplines such as sociology, psychology, political science, economics, or – for that matter – technology. Etymologically the term *management* derives from the Latin word *manus* (hand), meaning to handle, for instance, a tool or a piece of equipment. However when we talk today about management, or management studies as an academic field, we usually refer to studies on how to handle an organization (a firm, a public agency, a business, a production facility, a project, etc.). Specifically we refer to how to forecast, plan, structure, command, coordinate, and control the venture's resources (employees, finances, equipment, etc.) so that together they contribute to the venture's success. In most organizations, these resources must be managed under the condition of scarcity, that is to say, when the lack of resources prevents the satisfaction of all organizational needs and wishes. This scarcity means that managers in organizations have to make choices.

Industrial Management, which is an applied, multi-disciplinary sub-field in management studies, encompasses concepts, theories, models, and techniques that are found in, or originate from, other disciplines such as business administration, sociology, economics, and psychology. Thus, Industrial Management does not propose any different, or unique, models or theories. Instead, the difference between Industrial Management and other disciplines relates to its points of departure, emphasis, and objects of study.

Industrial Management typically takes the *operations* of industrial firms, and the *value-creating processes* associated with these operations, as the point of departure. While the related fields of general management and business administration focus on the management of resources and systems (e.g., financial accounting and reporting) of a company or other organization, Industrial Management primarily focuses on the processes of *innovation*, *production*, and *marketing* that together constitute the company's core operations. The aim of these processes is to *create value* for someone such as a customer, a client, or a user. However, value creation usually also implies

14

consumption of resources. Consequently, value creation must be efficient. If the value of goods and services created is greater than the value of resources consumed, a surplus (hopefully sustainable in the long term) results. Thus, development, improvement, and control of these value-creating processes are key issues in Industrial Management.

Another characteristic of Industrial Management as an academic field is its close association with technology, technological development, and engineering. The term "industry", which etymologically derives from the Latin word *industrial* that means "diligence, activity, and zeal", is today typically interpreted in an organizational context. The adjective "industrial" in the term Industrial Management has its historical roots in the technological developments of the Industrial Revolution of the 19th century. The discipline is closely associated with the emergence of the mass production system in manufacturing in the decades before and after the year 1900. Even today, more than 100 years later, mass production is the managerial basis of our material wealth and welfare.

It is not a coincidence that Industrial Management first emerged as an academic field in the mechanical engineering curricula, or that much of its early emphasis was the design and creation of efficient production processes in the manufacture of goods. However, in the last 100 years the academic field of Industrial Management has evolved in line with overall developments in industry, technology, and society.

Today, Industrial Management, which is a broad academic field, addresses many aspects of technological development, innovation, operations, and marketing that apply to more and more ventures and enterprises. Over the years, the field of Industrial Management has expanded to include a number of new sub-disciplines that address many different issues such as innovation, work science, strategy, industrial dynamics, organization theory, gender, supply chain management, logistics, project management, product development, and entrepreneurship. Thus, we conclude that Industrial Management is an academic field concerned with the issues of management and business that engineers need to master if they are to be successful in their professional work.

Three related subjects

1) Economics (in Swedish: *Nationalekonomi*)
Economics, which is the study of society's ability to create, organize, and use resources as a whole, is based on the assumption that all human needs and wishes cannot be satisfied simultaneously. People must make choices. Economics typically deals with groups of actors who make such choices in a stylized market. One sub-field of economics, which deals with the structure and evolution of industries and markets, is sometimes called Industrial Organization or Industrial Economics (not to be confused with Industrial Management).

2) Business Administration/Business Studies (in Swedish: *Företagsekonomi*)
Business Administration or Business Studies is the study of companies, their management, and their use of limited resources. Examples of sub-disciplines include financial accounting, corporate finance, management control, marketing, leadership, and organization theory.

3) Industrial Management (in Swedish: *Industriell ekonomi*)
Industrial Management is the study of business and management challenges that engineers face at work. Industrial Management concerns the value-creating processes of technologically intensive organizations such as, typically, the processes of innovation, operations, and industrial marketing.

Education programs in *Industrial Engineering and Management* are taught at most major engineering universities in Europe. These programs typically integrate studies of engineering and technology (about two-thirds of the curriculum), with courses in industrial management, etc. (about one-third of the curriculum).

1.2 Industrial Management – yesterday, today, tomorrow ...

Industrial Management emerged as a field during the second phase of the world's industrialization in the early 1900s. In these years, the premier challenge of industrial companies was to transform raw materials (e.g., ores, wood, and textile fibers) effectively to material goods in one or several steps at one or several companies. In Sweden the largest and most complex industrial companies at the time produced mechanical and electro-mechanical products such as separators, electrical power equipment, trains, telephone switches, and ball bearings. Other Swedish companies were involved with the operations of large technical systems such as telecommunications and power production and distribution. By international standards, Sweden is a small country; however, Sweden has a large number of major, multi-national industrial corporations. Many of these internationally well-known manufacturing companies were founded in the early 20ᵗʰ century. Some examples are ABB, Alva Laval, Atlas Copco, Electrolux, Ericsson, Sandvik, Scania, SKF, Telia, Vattenfall, and Volvo.

Over time, industrial companies have developed their management activities and their operations. Beginning with a value creation process based on the transformation of raw materials to physical products (goods), many companies now offer a broad array of goods, systems, and services. Although manufacturing remains their core competence, today many industrial companies employ their people in activities other than pure (and traditional) manufacturing activities. For example, many employees at contemporary industrial companies are involved in research and development, project management, systems integration, operations of complex systems, maintenance, and other services.

Many industrial companies today offer a number of services that support, complement, and sometimes even replace their traditional physical products. A well-known example is Rolls Royce, the British aircraft engine manufacturer, which, as early as the 1950s, offered customers maintenance service contracts at a fixed price per flying hour under the famous slogan "Power by the hour". Instead of purchasing the engines and paying for repairs as needed, customers signed a contract for a certain time period; in other words, they paid for engines that worked, not for engines as products or for engine repairs. Another example is the Swedish telecom company, Ericsson, which originally manufactured telephones and switches but today develops and produces complex telecommunications systems and sometimes also operates these systems for its customers. A third example is the American company, Apple, which, in addition to computers, mobile telephones, and tablets, offers an entire system of various services – all available in Apple's web-based, online stores. Thus, even if the term *industry* is still spelled as it was 100 years ago, today it has many more multi-faceted connotations than it had originally.

Usually, the term *industry* refers to business operations that have high technological applications (e.g., the operations of mining, manufacturing, pharmaceutical product development, or construction). Sometimes the term refers to sectors of the economy (e.g., the tourist industry, or the movie industry). However, when we characterize a specific business as industrial or *industrialized*, we intend a more specific meaning. The operations of businesses that are categorized as industrialized usually have four specific qualities: (1) they are large-scale, (2) they are systematically managed, (3) they have a high degree of task-specialization, and (4) they have a high degree of standardization. Thus, the concept "industrial" is associated, often implicitly, with mass production and repeatability, which in turn promotes automation and efficient utilization of machinery and equipment.

Two examples of operations that are not industrial (or industrialized) in the above sense are the handicrafts and the arts. The basic idea behind such operations is that their finished products are unique; if an artistic activity becomes too industrialized, the value of its products is typically greatly reduced. Other typically non-industrialized sectors include attorney offices, accounting firms, schools, and consumer services such as haircutting and massage. The commonality of these operations is that each customer is unique, and the operations are tailored to each customer.

Given this perspective on industrialization, it is possible to identify a number of business operations, which traditionally have not been regarded as industrial, as highly industrialized because of their large-scale, systematic, standardized, as well as specialized, processes. Today, a large number of service operations are organized by mass production principles similar to those of the traditional manufacturing companies. Corporations such as McDonald's, H&M, Zara, and Media Markt operate low cost businesses in which they produce and sell large volumes of their products. This *business model* requires, among other things, highly efficient operations, advanced logistics systems, convenient customer locations, standardized products, reliable suppliers, and major marketing programs.

In the same way, large-scale service suppliers work with systematically developing their processes so that they can stay competitive. For example, the Danish company, ISS, is one of the world's leading facility service companies with over 500,000 employees. The company's activities include facility management, cleaning, office services, and restaurant operations. Often, these activities have been outsourced to ISS by companies that previously performed them in-house. Another example is the Swedish company, Securitas, which is one of the world's foremost security firms with over 300,000 employees working with, among other things, safety controls, parking monitoring, alarm systems, and security advice. Using economies of scale, these companies can offer labor-intensive services based on standardized processes that have been systematically designed and tested.

Another important example of complex and labor-intensive service operations is healthcare, in which standardization is increasingly used to streamline patient flows, reduce waiting times, and improve the quality of healthcare services. Clinics, emergency departments, and even surgery departments more and more use practices that are managed following industrial principles.

In summary, we conclude that the concept of industry in Industrial Management, on the one hand, is closely associated with *technology-intensive* operations (i.e., technological innovation, systems development, advanced calculations, engineering design, complex machinery, etc.). Typically, engineers are engaged in these types of activities.

On the other hand, we also conclude that the boundaries of Industrial Management are continually changing and evolving. The concepts, tools, and techniques originally used in Industrial Management are now used in a great variety of businesses. At the same time that many traditional manufacturing companies are developing their operations by offering a greater share of service content, many traditional service companies are struggling to industrialize their operations so that they can grow and become more efficient. The academic field of Industrial Management deals with both situations.

Industrial Management – its academic history in Sweden

- 1912 – Mr Erik Forsberg, senior engineer at Separator (today Alfa Laval), is employed as a specialist teacher in Industrial Economics and Management at KTH Royal Institute of Technology
- 1939 – Mr Tarras Sällfors joins KTH as Sweden's first Professor of Industrial Economics and Management
- 1969 – Linköping University inaugurates Sweden's first comprehensive MSc Program in Industrial Engineering and Management (the I-program)
- 1972 – Mr Sven Åke Johansson earns Sweden's first PhD in Industrial Economics and Management at KTH
- 1980s – Chalmers University of Technology and Luleå University of Technology introduce MSc Programs in Industrial Engineering and Management
- 1990s – KTH and Lund University introduce MSc Programs in Industrial Engineering and Management
- 2010 and forward – Industrial Engineering and Management has become one of the most popular fields in Swedish engineering education and is offered (at basic and/or advanced levels) at the engineering faculties at most Swedish (as well as European) universities

1.3 Value creation and the economic cycle of an industrial venture

Industrial Management revolves around a company's operations – that is, it begins with what the company does, the content of its business, and its value-creating processes. Chapter 2 discusses these ideas in detail. The operations of the company create a financial system, a kind of economic cycle, in which revenues are generated from the value creation of goods and services. Furthermore, the operations usually require an organization of human resources, raw materials, production facilities, offices, warehouses, and machinery/equipment of all kinds, which means that the company must have enough financial strength to support these activities. This book explains many of the business and management concepts, tools, and techniques used

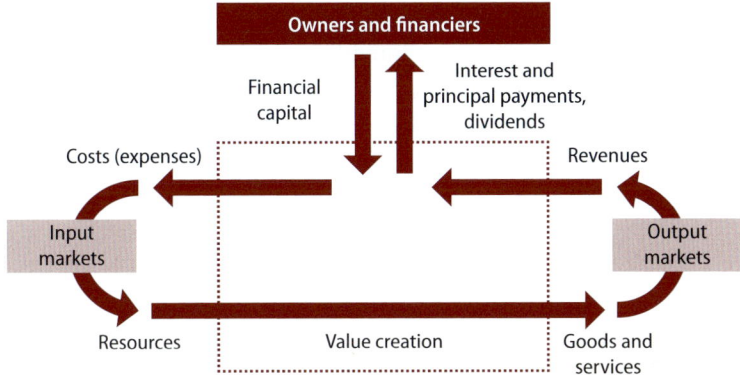

Figure 1.1 A company's economic cycle.

in organizing, planning, and controlling these operations. For example, the book describes product cost estimations and cost allocations, decisions on capital investments, financial accounting, and management control.

Figure 1.1 presents a stylized economic cycle for a traditional manufacturing company. A company requires *financial capital* (typically cash) provided by different types of *financiers* such as company owners, banks, or other investors. This capital finances the company's initial activities. The capital invested by the owners constitutes the initial *owners' equity* in the company. Bank loans are liabilities requiring *interest* and *principal payments* under specified terms. In the long run, the investors require a financial return on their investment. This decision of how to finance (capitalize) a company at its founding is a strategic decision that often has long-term implications for the business.

The company uses its initial financial capital to acquire *resources* such as personnel, raw materials, office/factory space, and other items including the machinery/equipment needed for its activities. The company acquires these resources on various *input markets*. For example, when various alternatives are available for the purchase of *raw materials*, the company must calculate and compare prices, volumes, etc. of different alternatives. Once the *value-creating* processes begin, the company must design a system for the allocation of its various *costs* to the various products and services so that it can set appropriate sales prices. Finally, the produced products are sold on the *output markets* at prices that are intended to return a profit.

Some resources, such as machinery and equipment, are essential components of the company's operations system. These items are often expensive and long-lived. Therefore, they influence both the extent of the company's operations and its financial outlook – far into the future. Purchases of machinery and equipment are called *capital investments*. The management and control of such capital investments are often critical for a company's growth and development.

To manage its activities, a company must have a relevant *organization structure*. This structure is essential so that the company can manage and monitor its management procedures and processes effectively. Companies usually draw organization charts that depict reporting and authority lines. In addition, the organization structure describes the division of labor, defines responsibilities, states job specifications, and specifies managerial authority.

The company's products – goods and/or services – are sold on *output markets* and generate, if successful, sufficient *revenues* to pay for salaries, raw materials, equipment, interest on loans, etc. Each year the company closes its financial accounts (bookkeeping records) so profit (or loss) can be calculated, and corporate taxes and other fees can be paid to the government. If a surplus remains after taxes and fees have been paid, the remainder can either be divided among the owners as cash dividends or be retained in the company as part of owners' equity. The economic cycle is closed.

Engineers need to understand this economic cycle and how its dynamics affect (and are affected by) the various engineering activities of the firm such as the development of new technology, operations management, product development, project management, systems engineering, and marketing. Although many engineers are only active in one or two of the specific steps in the economic cycle, they can greatly benefit from an understanding of how the cycle works. The economic cycle is closely related to a number of significant concepts, tools, and techniques used in contemporary technologybased businesses. It is essential to understand these concepts, tools, and techniques in order to create, develop, implement, and sustain competitive value creation.

CHAPTER 2

Value creation as the point of departure

Industrial operations (i.e., industrial activities) can be described in many different ways. One place to start is by focusing on the activities of individual companies as they develop, produce, and market their products (i.e., their goods and services). We can also study differences among companies as far as their products and production processes. In this chapter, as we discuss industrial operations, we examine how the company creates and sells its products, how it consumes resources in those processes, and how it generates revenue from those processes. Three fundamental concepts are in focus – *value proposition, value creation,* and *value capture*. In essence, we are describing the company's *business model*, which we need to analyze if we are to understand how a company's various activities interact and influence each other.

2.1 Value propositions: goods and services

Industrial companies create value by developing, producing, and marketing various *products*. In everyday usage, when we speak of products we are referring to physical objects. However, more specifically, products can be *goods* (tangible products) or *services* (intangible products). Traditionally, Industrial Management is concerned with the production of goods; however, with time, the production of services has become increasingly important. To avoid the traditional understanding that a product is only a tangible object, people today often talk about *value propositions* instead. This concept refers to the value of a company's offerings to its customers, whether goods or services.

Various types of value propositions

Some companies sell their products directly to consumers, often called Business-to-Consumer (B2C). Other companies sell their products to other companies, often called Business-to-Business (B2B). See Table 2.1. This is a simple classification, but the difference is not always clear-cut. Some products are sold to both consumers and producers. However, the customer type is often of critical importance for the company's strategy and operations.

Goods and services differ in several ways. Tangible products (goods) can be developed, produced, stored, and delivered. Frequently, the development, production, and storage of goods occur before the customer is involved or even identified. In addition, it is often possible to assess and measure the quality of tangible products in precise and objective ways. It is a different matter for intangible products (services). Because of their intangibility, pure services are

Table 2.1 Examples of various product types.

Business-to-Consumer (B2C) goods	Business-to-Business (B2B) goods	Business-to-Consumer (B2C) services	Business-to-Business (B2B) services
Books	Pulp	Education	Vocational training
Automobiles	Trucks	Auto repair	Maintenance
Toothpaste	Plastic granule	Dental services	Corporate health plan
Mobile telephones	Telecom systems	Telecom services	IT support
Medicines	Laboratory equipment	Health care	Corporate health plan
Washers	Steel	Laundry	Catering
Laptops	Electronic components	Electronic payments	Auditing
Houses	Factories	Cleaning	Facility services
Dinner table	Office furniture	Gardening	Interior design
Compact Discs	Recording equipment	Streaming media	Technical consulting

produced and consumed simultaneously in a direct, close producer-consumer relationship. Moreover, typical services cannot be stored and delivered (note, however, that storage and delivery are also services). Evaluating the quality of services is often a highly subjective process.

In practice, however, most products are a mixture of goods and services. Many tangible products, such as advanced machinery, are sold with service provisions. On the other hand, mobile telephone subscriptions (services) sometimes come with new telephones (goods) "for free". The installation of large computerized business systems (ERP – Enterprise Resource Planning – systems) requires considerable work to adapt them to the purchasers' specifications. In short, when we, as consumers, purchase a tangible product, the chain of operations behind the product consists of a number of pure (intangible) services. For example, if we purchase a computer via the Internet, this is a purchase of a physical product. The computer and its components were manufactured and assembled at various factories in the world. In addition, various services actually make up a large share of the activities resulting in the delivery of the computer. These activities include, for instance, recording the order, installing the software, configuring the operating system, and shipping the computer. See Figure 2.1.

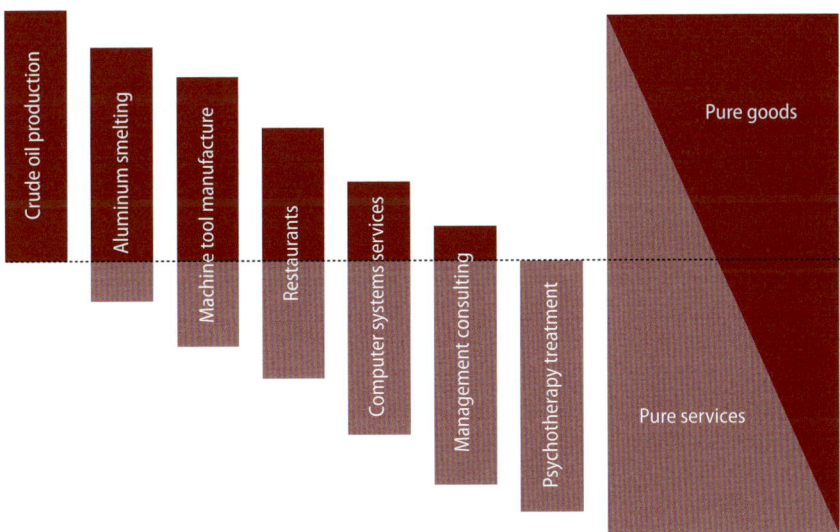

Figure 2.1 Value propositions: Most are a mixture of manufacturing and service operations.

The various services that industrial companies offer their customers are of two general types. First, there are services that *complement* the company's physical products by *smoothing* (facilitating) the sale and use of the product (e.g., by offering a service) or by *adapting* the products to the customer's business so they work properly in combination (e.g., by customization of a new IT system). Second, there are services in which the company *substitutes* services for the physical products. That is, instead of purchasing a product, the customer leases the product, subscribes for its use, or specifically purchases the functionalities the physical product provides. For example, instead of purchasing and operating their IT systems, many companies purchase IT operations services from external suppliers. Similarly, many companies purchase a set of specific copier and printing services from a supplier instead of purchasing a copier and a printer. The suppliers then provide maintenance and repair services. In fact, many industrial companies today are developing such business models as a way to increase the value of their offerings and to promote long-term and closer customer relationships. Moreover, for some customers, making regular, fixed, and smaller payments for services over a longer period is preferable to financing a large purchase of expensive capital equipment. See Table 2.2.

Table 2.2 Functions of industrial services (based on Cusumano et al., 2015).

Complementary with physical products		Replacement of physical products
Smoothing	*Adapting*	*Substituting*
Services that simplify sales or use of a product but do not significantly affect the product's functionality.	Services that adapt the product, improve the product's functionality, or help the customer develop new ways to use it.	Services that replace the purchase of the product.
	Tightly linked to the product. Requires close cooperation with the customer.	
Examples: • Financial services • Warranties • Insurance • Technical support • Training • Operations support	Examples: • Custom tailoring • Systems integration • Technical updates • Development of a customer's business processes	Examples: • Computer/IT services • Internet data storage • Rentals and leases • Functional sales "Power by the hour"

Value propositions and company strategy

How a company formulates its value propositions is closely associated with its *business concept* (sometimes the term *business idea* is used) and its *strategy*. The company's business concept reflects its overall goals, the reason for its existence, and the owners' vision. A good business concept also reflects more aspirations for the company's activities than those reflected by its financial goals. The company's *strategy* reflects the company's plans for how it intends to realize its business concept – that is, how the company will provide value to customers, how it will compete with other companies, and how it will develop its activities on a long-term basis.

Designing the value proposition is a strategic decision that influences all the company's activities – for example, the market segment in focus, required competences and skills, equipment needed, and, not least, managerial, financial, and operational plans. Regardless of whether the company offers goods and/or services, there are three principal alternative, competitive strategies to choose from. See Figure 2.2.

1. *Competitiveness by product leadership*: The value proposition is based on superior products (i.e., goods and/or services). By the high quality of the products' properties and the sophistication of their functions, the company creates a distinctive brand. To maintain this level of quality, the company must invest heavily in innovation and product development. The success of the business depends on its ability to command relatively high (premium) prices for its products.

2. *Competitiveness by operational excellence*: The value proposition is based on low prices. When operational processes are efficient and products are standardized, products can be manufactured at low cost and in high volume. Sales prices are low (in relation to the product quality offered). The success of the business depends on its ability to maintain low production costs.

3. *Competitiveness by closeness to customer*: A company manufactures tailor-made products for its customers. Unique products are produced to order. Customer loyalty and relationships are very important. The success of the business depends on its ability to maintain operational flexibility in meeting specific customer requirements.

In practice, of course, many companies combine these strategies in one way or another. However, it is important to remember that there are inherent, conflicting logics among these three strategies. High operational efficiency is usually incompatible with high flexibility as well as with highly customized production. Similarly, it is very difficult to develop sophisticated product properties, premier features, high quality functionalities, and at the same time offer these products at low prices. Thus, a company's *value proposition* and its overall strategy must match the company's three major *value-creating processes*: the processes of innovation, operations, and marketing. We will return to these processes later in the chapter.

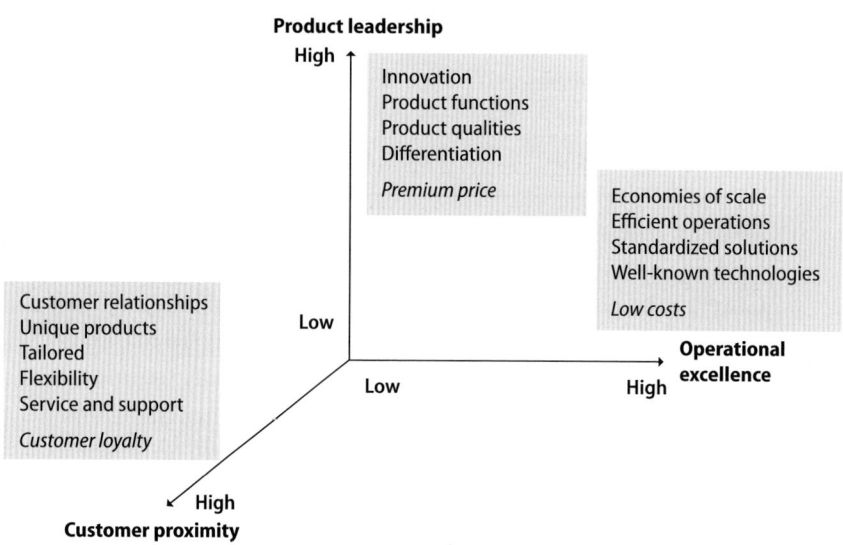

Figure 2.2 Three competitive strategies.

2.2 Industrial value creation

The company's transformation of resources

A classic model of *value creation* in industry describes industrial operations as the *transformation* of various *resources* (inputs) to *products* (outputs). A company's *value added* is the difference between the economic value of its output (sales price for finished products) and the economic value of its input (cost of the resources that make up the products).

The value-creating process is different for different businesses. In the pure *manufacturing company,* various inputs are transformed to outputs; that is, to the physical products (goods) sold to customers. See Figure 2.3. Inputs may be raw materials, semi-finished goods, and various other components. The resource transformation process also requires production plants, technical equipment, the knowledge and skills of personnel, and (sometimes) external contractors.

In the pure *service company,* the transformation and logic of the value creation process are different. The customer is often directly involved in the service operations (e.g., airplane passengers). Sometimes the service involves changes to the customer's own operations or properties (e.g., auto repair services, computer system installations, or freight transport). Or the service may involve knowledge development (e.g., training, consulting, data collection, or statistical analysis) for the customer.

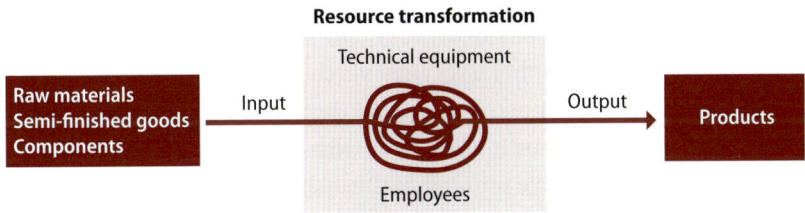

Figure 2.3 Value creation in manufacturing operations.

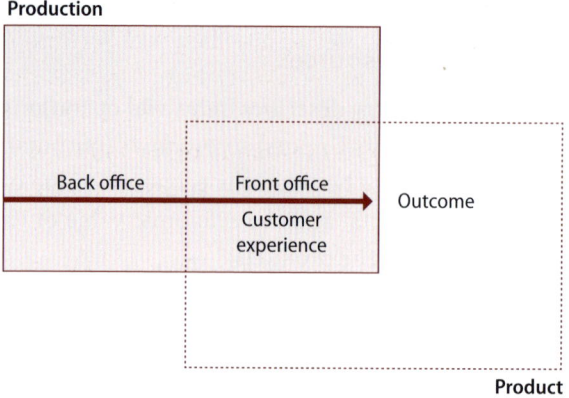

Figure 2.4 The product as the sum of the outcome and the customer experience (based on Johnston & Clark, 2008).

The degree of customer interaction

In most cases, value creation results from a mixture of goods and service production. Figure 2.4 presents a simplified illustration of this resource transformation process. The arrow in the upper box represents operations. The lower box represents the customer experience. As Figure 2.4 illustrates, some areas of the two boxes overlap although other areas do not. This overlap differs according to the type of operations. See Figure 2.5.

Normally, customers are not involved in traditional product manufacturing (e.g., the manufacture of automobiles or computers). Thus, there is no overlap of the two boxes. The same is true of certain kinds of services (e.g., delivery of mobile telephone services via the Internet). Similarly, banks and travel bureaus conduct many of their activities in the back office. Their customers have no contact or involvement with such activities.

Typically, however, a large number of the operations in the service sector require direct contact with customers. Examples are the front office activities of banks and travel bureaus in which the customer experience is part of the service product purchased. In such cases, the value proposition is the sum of the outcome (end result) and the customer experience.

Traditional production activities of manufacturers and back office activities of service companies often have many similarities. Generally, value creation is independent of the customer, who only evaluates the final outcome. However, when the two boxes overlap, the larger the overlap, the more direct contact between customers and operations, and consequently the more relevant the customer experience to the performance of the operations. At a restaurant, for example, the quality of the food (from the back office) is fundamental. However, from the customer's perspective, the restaurant's service, atmosphere, and hospitality (in the front office) are at least as important. The same situation exists for the installation of technical systems or for the provision of technical consulting services. Naturally, the technical outcome is the most important aspect, but the customer experience with the technicians and consultants may influence the outcome either positively or negatively. Were the technicians and consultants easy to work with? Did they meet the time schedules? Were they knowledgeable and reliable?

In addition, there are service-based value propositions in which the very process constitutes the product, and where there are almost no back office activities. See Figure 2.5, in which the three boxes merge. Examples of such services are various therapies, for instance, massage. However, such activities are seldom viewed as industrial operations.

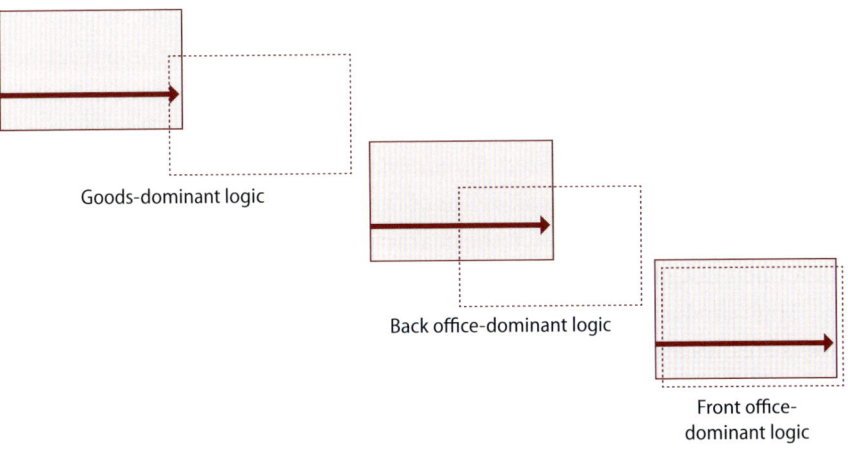

Goods-dominant logic

Back office-dominant logic

Front office-dominant logic

Figure 2.5 Various degrees of customer involvement in value creation.

Based on the customer interaction, the value-creating process can be divided into three principal categories (we will return to the strategic consequences of the division later in the chapter):

1. *Goods-dominant logic:* Value creation is independent of interactions with the customers.
2. *Back office-dominant logic:* The main value creation is independent of interactions with the customers, but interactions are still central parts of the process (often at the beginning or the end of the process).
3. *Front office-dominant logic:* The main value creation occurs in direct contact with the customers; most of the value creation involves the customer experience.

Volume or variety

Among the many factors that influence a company's resource transformation process, the volume of operations is one of the most important. High production volume and high turnover (sales) volume create the conditions for low production costs, which in turn are among the most important competition factors. The concept of *economies of scale* is the idea that with high production volume, a company can achieve low cost per unit based on efficient utilization of labor, equipment, production plants, etc.

Achieving economies of scale, however, requires that the operations processes are predictable and repeatable. Thus, these processes must be designed and managed *systematically* so that all components, sub-processes, and activities are standardized. *Standardization* creates opportunities for the *specialization* of employees as well as of the equipment, which can lead to still greater economies of scale (cf. the characteristics of Industrial Management described in Chapter 1).

High production volume is often capital intensive. High volume in operations enables the purchase of expensive, specialized equipment, which then leads to the possibility of increased operational efficiency. On the other hand, competition through high production volume and low costs usually requires major investments in specialized equipment and in expensive production facilities.

High production volume, thus, is tightly linked with company strategy and competition by *operational excellence*. See Section 2.1. There are many examples of operations in which the ability to maintain high production volume is a prerequisite for survival. Classic examples are the process industries such as oil and gas, steel, pulp and paper, chemicals, and food. These operations are largely dependent on high production volume. Other examples are the manufacture of vacuum cleaners, cars, computers, and electronic components as well as the service operations of freight companies, fast food restaurants, movie theaters, and retail stores (e.g., IKEA, H&M, Primark, and Walmart). Perhaps the most well-known example of high volume service operations is the hamburger chain McDonald's Corporation. McDonald's, which is the largest restaurant company in the world, serves almost 70,000,000 customers daily.

There is a trade-off, however, in the standardization and specialization of high volume production: operations become less flexible and the goods and services provided can only exist within a limited product range. Moreover, production is often expensive and inflexible. If market demand changes, or if some customers ask for products with special features, it is usually very difficult to adapt the standardized products.

Businesses with high variation in their value propositions (the strategy of competition by customer closeness described in Section 2.1) must be flexible. This usually means less predictability and specialization in the operations, and, as a result, processes with significantly lower levels of repeatability. The results are higher unit costs for goods and services. Therefore, such companies must create value that justifies higher sales prices than those of companies with larger production volumes.

For example, compare a fine-dining restaurant with a fast food or take-out restaurant. The fine-dining restaurant usually offers many more choices on the menu, all of which are prepared by skilled, creative, and experienced chefs. The meals, which also take longer to prepare, are also more expensive. Therefore the fine-dining restaurant must justify its higher prices by the quality of its food, the competence of its servers, and its atmosphere.

This restaurant example focuses on the classic juxtaposition of high volume operations and high variety. High volume operations require repeatability, specialization, and standardization. High variety that meets customers' demands requires high flexibility and skilled competence in

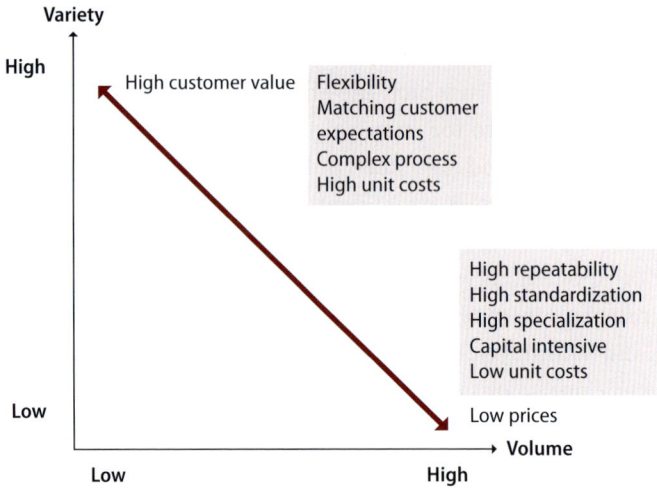

Figure 2.6 The classic compromise between production volume and product variety (based on Hayes & Wheelwright, 1979).

complex production processes. Of course, there is a middle ground between these two extremes of volume and variety. Figure 2.6 presents the classic (compromised) relationship between them.

In practice, many companies in various ways must compromise between high volume and high variety. For example, some companies adopt high variety when they offer their customers ready-made packages of goods and services. Other companies try to join the two processes, for example, by using mass-produced components or sub-processes that are combined in a product or service. From the customer's perspective, the product or the service appears unique.

The inherent features of production flow

Another dimension of the value creation process is the logic of the production flow. This flow is usually tightly linked with the design and the components of the products (goods and services). Using a simplification, we can categorize these logics into four types: *convergent flow, divergent flow, hourglass flow,* and *T-shaped flow.*

34

Convergent flow production is common in the traditional manufacturing sector. In such production flows, the inputs of raw materials and a large number of other components are converted to the outputs of fewer products, some of which are very complex. The production process for such products is also complex. Planning, coordination, assembly, and logistics are required for many components that must be designed and assembled so they function together. Both high volume and high variety are characteristics of this type of production. Often, very complicated products are produced in small volumes, although there can also be large-scale production volume of components and sub-systems.

An example of convergent flow production is the manufacture of Boeing Company's Dreamliner, which consists of more than five million components per plane. See Figure 2.7. Another example is a software development project that requires coordination by a great many specialists with various technical and managerial competences.

In *divergent flow* production, the flow is the opposite of convergent flow production. In this production process, a few inputs such as raw materials and other components are converted to many different end products. The logic of the flow is based on high volume and highly specialized processes intended to achieve economies of scale. An example of a company that uses

Figure 2.7 Example of a product from convergent flow production.
Source: mirounga/shutterstock.com.

Figure 2.8 Example of a product from divergent flow production.
Source: wang wentong/shutterstock.com.

divergent flow production is SSAB, which produces many different grades of steel from, in principle, an alloy of iron and carbon. Another example is an oil refinery that converts crude oil to a great many products – from gas to asphalt. See Figure 2.8.

In some production flows, variation in the value proposition is achieved by creating specific product designs just before product delivery to customers (the *T-shaped flow* production). An example is a company that manufactures aluminum cans for the brewing industry. In principle, the company produces only one type of can. However, in the last phase of the production flow, the company produces a great many variations in the can design. Other companies use the same logic. For example, clothing companies which manufacture T-shirts from one pattern and one fabric can, in the final production phase, use different colors and different print designs. With this flexibility, they can produce at high volume while at the same time keeping up with changing fashion trends.

Other companies' production flows instead resemble hourglasses. *Hourglass flow* production combines the characteristics of the convergent and divergent flows. An example is the company whose products are *modularized*. In this flow, standardized modules (high volume) are manufactured from a large number of components. These modules, which can be combined in many different ways, result in a wide range of customer products (high variety). Well-known examples of modularized products are Scania trucks. For decades, Scania has systematically developed a number of different variations of cabs, gearboxes, engines, frame components, etc. These items' modules can be combined in many different ways to suit the customers. See Figure 2.9.

IT system providers have a similar strategy. By offering various programs and system modules that can be combined in different ways, these companies standardize the development processes for each module at the same time that they meet the customers' particular requirements for each system.

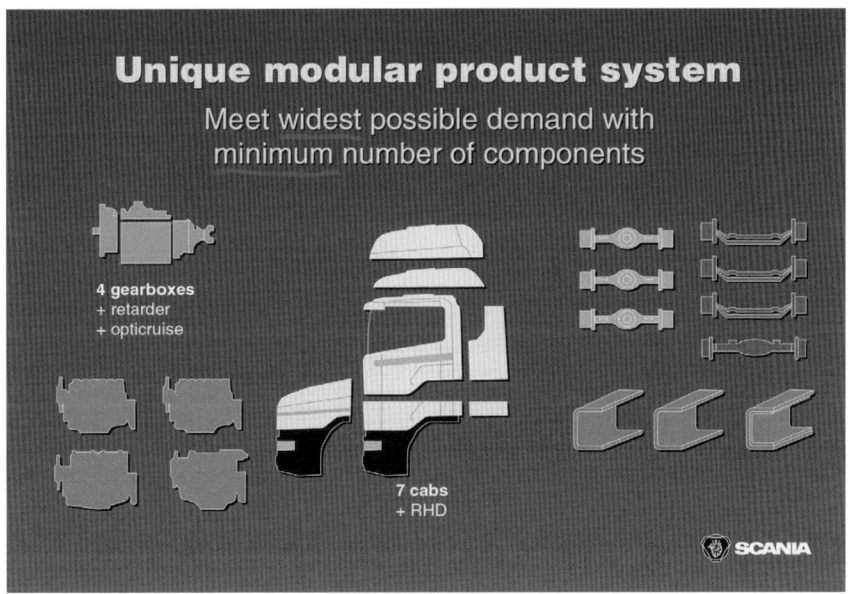

Figure 2.9 Modularization is a way to combine convergent and divergent flow production. Source: Scania.

The company's processes and functions

To this point we have discussed the company's value creation from the point of view of operations. However, a company's innovation and marketing processes are just as important. *Innovation, operations,* and *marketing* are the three *value-creating processes* that in combination result in the company's total value creation. *Innovation* deals with development of new ideas in the creation of value propositions (e.g., new products). *Operations* deal with the efficiency of realizing value propositions (e.g., production). *Marketing* deals with bringing value propositions to the market and with managing customers' responses.

Most companies (if they are not too small) are structured into various functional departments (or units or divisions), each with responsibility for some phase of the three processes. These departments may have different names and may be organized in different ways at different companies. However, typically, one department type deals with research and/or product development (innovation), one department type with manufacturing, production, or service operations (operations), and one department type with sales, marketing, or external communications (marketing).

As Figure 2.10 illustrates, none of the three processes is confined to only one department type. The processes spread across the departments as the company's various activities are performed. For example, *operations* are influenced by activities in departments other than the manufacturing department. The sales department (i.e., *the marketing process*) influences

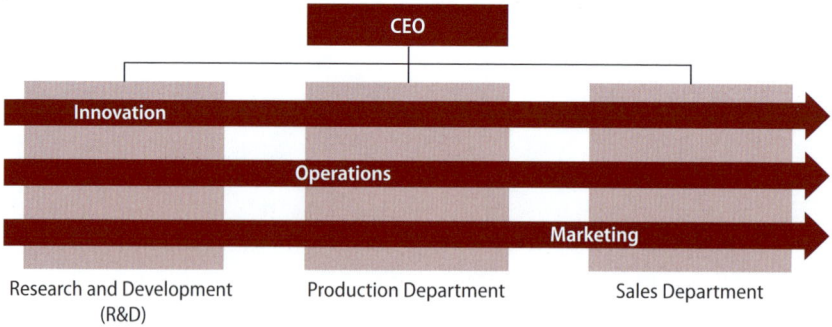

Figure 2.10 Three value-creating processes – three typical department types in the industrial company.

operations with its transmittals of customer demands for delivery times. Operations are further influenced by *innovation* in the way the products are designed and in the way they are intended to function. Moreover, *marketing* deals with more than the company's market-related activities. Marketing activities and strategies also influence the design of products, their performance, and their quality (see, for example, Intel's slogan "Intel inside" or clothes advertised as made from Gore-Tex fabrics).

In addition, employees may be involved in two or more of these processes although they may not always be aware of this fact. For example, the front office employees are often involved in a mixture of operations and marketing processes. Other examples are the service technicians, caretakers, transport personnel, and consultants (in the operations process) who work at customer locations and influence customers' perceptions of the company (in the marketing process). Furthermore, an advertising agency's operations consist of marketing activities, a consulting firm's operations consist of consulting work, and a research laboratory's operations consist of innovative research activities.

These three processes – innovation, operations, and marketing – span the activities of the entire company – from purchase of raw materials to the sale of the finished products (goods and services). If described individually, these processes present slightly different pictures of the company. Innovation emphasizes the product or service and its qualities (value proposition); operations emphasize resource transformation (value creation); and marketing emphasizes the customer and customer product perception (value capture). It should not be forgotten, however, that these three value-creating processes interact and influence one another.

2.3 Value capture

To survive in the long term, a company must retain some of the value it creates for its customers in the form of revenues that generate profit. This is referred to as *value capture*. The way this is done, known as the *revenue model,* is closely associated with the value proposition the company offers its customers.

Most traditional revenue models are built around *unit sales* – that is, the company receives payment for each unit sold for an amount that exceeds the costs to produce that unit. The difference is the profit margin. Because cost-based pricing is easy for everyone to understand, companies tend to price their goods and services on a per unit basis. Thus, sales prices reflect the cost of raw materials, labor, and overhead, plus a profit. However, the price customers are willing to pay for a product cannot be objectively determined. Market conditions set sales prices based on rather vague ideas about the value of the product to the customer, the price competitors set for an equivalent product, and the alternative product the customer could have purchased with the same amount of money.

There are, nevertheless, a number of other ways to generate revenue and to capture some portion of the value created for customers. One way is by *additional sales* in what is often called the aftermarket. In such sales, the company sells additional equipment, spare parts, maintenance, or other product-specific services to customers who have become somewhat dependent on the company because of their original product purchases (from the supplier perspective, this is often called the company's "installed base"). Typical examples are car retailers that often sell original spare parts and accessories, or car repair shops dedicated to specific automotive brands.

One common version of this revenue model is sometimes called "Razors and Blades". The name comes from the now famous strategy used by Gillette for its brand of men's safety razors. The razors were cheap, but the replacement blades were relatively costly. In a nutshell, this strategy tempts consumers to buy an inexpensive product that requires more expensive *consumables.* This revenue model is commonly used with both consumer and producer goods. For example, compare the price of a desk printer with the price of its ink cartridges, which, of course, are specifically designed for each printer model.

Another version of this revenue model, which is often associated with luxury products, is to sell *accessory products* and add-on services under the same brand even though these products/services have no direct connection to the original products. For example, the luxury automobile manufacturers Porsche and Ferrari also sell clothes collections under their brands. The idea is to strengthen customer loyalty through brand identification as well as to generate significant additional revenues.

A third type of revenue model is to *rent or lease* products instead of selling them. Although the company retains ownership of the product, the customer pays for the right to use the product. This model is often combined with various kinds of maintenance and warranty contracts that guarantee a certain level of product performance.

A fourth revenue model is *licensing*. In this model, the customer purchases the right to manufacture or sell a product, to use a technology, or to use a company-owned intangible asset (e.g., patents and trademarks). A prime example of this is the Danish brewery, Carlsberg, which, among other things, licenses the production of its Sol beer to other breweries around the world. Other examples are the famous fashion companies such as Armani, Hugo Boss, and Prada that increase their revenues by licensing their brands to cosmetics, perfumes, and other beauty products manufactured and sold by L'Oréal and others.

There are several other types of revenue models. In one model, the supplier company receives revenue from the customer's *product use*. One example is pay-per-view in which customers pay to watch events via private broadcasting. Another model is *freemium* (free + premium) in which customers pay for digital products such as software, games, and other Internet services. Adobe and Spotify, among others, use this revenue model in which the customer receives the basic product free of charge but typically must pay a fee to obtain access to some of the product's key features, to improved functionality, or to related Internet services.

Furthermore, in *affiliated marketing*, which, for example, Google and Facebook use, the product (goods or service) is offered free of charge to users while the company receives revenue from the sale of advertisements targeted at particular users. Publishers of free newspapers, commercial radio/television companies, and telecommunication companies (in the latter case, free telephony) also use this revenue model.

While this list of revenue models is not all-inclusive, it does reveal the breadth, ingenuity, and diversity of such models. The point is that these models are very often as strategically important as the company's product design. Development of new advanced technologies, including mobile payment systems, keyword recognition, and big data analysis capabilities, make new revenue models increasingly important.

In summary, how a company generates revenue is a key component of its business strategy. The different components of the company's business model (i.e., the value proposition, the value creation, and the value capture) form an integrated whole that must be consistent and function together.

2.4 Efficiency, effectiveness, productivity, and profitability

There are several ways to determine if company operations are "good" or "poor", or if they are managed "well" or "poorly". We often talk about the efficiency, effectiveness, productivity, and profitability of an activity or an organization.

The terms efficiency and effectiveness are closely related. *Efficiency* refers to how well the company's internal operations and activities perform the transformation of various resources to finished products – that is, whether this transformation is performed in the right way. This (internal) efficiency can be measured in different ways, including by the number of products manufactured per a specific unit of time, by the number of labor hours per product, or by the amount of energy consumed per product.

Effectiveness refers to the company's capability to meet external demands such as those from shareholders and customers. Does the company produce products that satisfy its customers? Consequently, a company's environment determines whether an activity is effective. Thus, efficiency relates to how well an activity is performed, while effectiveness refers to the usefulness of the activity.

Efficiency has no value if the effectiveness is poor. In other words, efficiency deals with doing things right, and effectiveness deals with doing the right things. However, there is a constant dilemma in that high efficiency in relation to specific objectives is often accomplished at the cost of less flexibility and increased difficulties in adapting to external changes, for instance, in customer demand. Consequently, a successful business must be "ambidextrous" (i.e., both efficient and effective at the same time).

Efficiency, effectiveness, and productivity

$$\text{Efficiency (``Doing things right'')} = \frac{\text{amount of output}}{\text{amount of input}}$$

Effectiveness ("Doing the right things") = degree of goal achievement

$$\text{Productivity} = \frac{\text{value of output}}{\text{amount of input}}$$

Another commonly used concept in economics and business is *productivity*. There are several productivity measures, but they all relate produced value (output) to consumed resources (input). Examples are labor productivity (value produced per labor hour), capital productivity (value produced per monetary unit), or total productivity (value produced in relation to all resources consumed during the resource transformation process).

There are two additional concepts of importance that are often used in performance measurement and financial control: *profit* and *profitability*. These are financial concepts and relate to efficiency, effectiveness, and productivity only indirectly. In a market economy, the fundamental assumption is that only effective companies have a chance of long-term survival and profitability. However, this does not mean that operations that are efficient and effective are automatically profitable.

Profitability is about financial effectiveness. It is a measure of whether the company's *profit* satisfies the market's or the owners' required return on their financial investments. Does a firm produce a reasonable profit given the amount of capital invested? There are many measures of profitability (as described in Chapter 7), but they have one commonality: the *profit* from the business (i.e., revenues less costs) is assessed in relation to the financial capital invested (such as facilities, equipment, cash, etc.). From the owners' perspective, a company with high efficiency, satisfied customers, and profit may not necessarily be profitable (enough). The owners may require a larger return on their investment than the company's profit produces, and there may be investment alternatives that provide a higher return than the investment in question.

Profit and profitability

Profit = Revenues – Costs

$$\text{Profitability} = \frac{\text{Profit}}{\text{Invested Financial Capital}}$$

How do we determine if a company is efficient, effective, and productive enough to be profitable? The answer to this question requires continual control and follow-up of the company's activities so that gradual changes are noted. This means using measurements (e.g., labor productivity) and comparing the results of these measurements to company goals and objectives. It also means developing the appropriate analytical metrics that are key elements in industrial management.

CHAPTER 3

Organizing operations

In many small companies, the same individual (the owner-entrepreneur) performs all work activities. Sweden, for example, has many small companies in which the owner alone purchases all materials and equipment; creates, manufactures, and sells all the products; deals with all the customers; and takes care of all the administrative activities including accounting, etc. If the company grows, and the owner hires employees, then others in the company share responsibility for these various activities. At this point, the company requires a formal organization structure.

3.1 Organizing an organization

The word *organization* has at least two meanings: unit or association, and structure (cf. *organization structure*). Typically, a formal organization unit (as differentiated from a spontaneous, informal group of persons) has four characteristics:

- it is an association of individuals
- the individuals perform different work tasks
- the work tasks are coordinated
- the purpose is to achieve a common objective.

Formal organizations take many different forms, depending on their members' various goals and various interests. Examples of organizations are the following: *member organizations* (e.g., sports clubs or unions) that serve the interests of their members; *business organizations* (e.g., private and public companies) that benefit their owners/shareholders; *service organizations* (e.g., healthcare centers) that serve users; and *societal organizations* (e.g., police forces and judicial groups) that serve society as a whole. How an organization prioritizes its activities depends on its organizational goal, and thus on how it manages and controls these activities.

Organizing

Specialization and *coordination* are two key concepts in the discussion of the concept of organization, that is, *organization structure*. Organizing typically means creating some kind of structure. In other words, organizing involves the attempt to create an enduring pattern of behavior and action among the employees that promotes effectiveness, efficiency, and cooperation.

A commonly accepted definition of the concept of organization structure is the following:

> all the means an organization uses to divide and coordinate the work with the intent of creating stable behavioral patterns.

If a small company with five employees, for example, employs an accountant to manage the company's financial accounting, the owner must define the accountant's work and responsibilities. How shall the job be designed? Other than managing the bookkeeping and accounting, should the accountant also manage company correspondence, make various calculations, deal with customers, and safeguard the archives? The accountant must know the limits of his or her decision-making authority – that is, can the accountant make decisions or must he or she always consult with the owner (e.g., for the purchase of new bookkeeping software or to make a change in the company's archives system)?

So long as a company employs only a few people, most of the division of labor among the employees is relatively simple and can be handled informally. However, as the company grows and hires more people, a much more elaborate and explicit division of labor is needed. Employees no longer know all the other employees or know what these other employees do, particularly because the company may now have employees at different locations. Individual employees now lack the knowledge and competences to solve every problem that the company encounters.

Therefore, a *division of labor* is essential. This means that different people in the organization allocate their attention to different issues and problems. Through this *specialization*, the various employees develop their knowledge of, and competences in, particular areas. Their expertise in these areas thus enables the company to develop its operations and become more effective and efficient.

At the same time that jobs are specialized, *coordination* is required among various parts of the organization so that everyone works towards the same goals. It is coordination that makes an organization more than just a collection of scattered individuals. The idea behind organizing is that it is easier to achieve an objective by collective group effort than by many individual efforts.

Organized action requires coordination

Coordination and control of activities and behavior in larger companies are best accomplished by creating organization rules and roles and by establishing norms and values. *Rules* standardize certain behaviors and actions, such as how to prepare financial reports, how to submit vacation requests, how to implement changes in product design, or who has the authority to purchase office equipment. Some rules are formalized as written *instructions*, decisions, or plans while others are implicit and manifested as *routines* applied during the daily operations.

However, you cannot depend only on rules, because rule-makers cannot predict all situations and required work tasks in advance. Even if that were possible, it would be impossible for people to remember all the rules for every situation. Therefore, all organizations also have more or less standardized organization *roles*. Department head, project manager, senior engineer, and marketing executive are examples of such roles. A role defines a broader framework for action than specific rules. At the same time, a role is limited by the various expectations and values associated with it. The greater the unpredictability in an organization's operations, usually the greater importance the roles have compared to the rules.

An organization also coordinates and controls employee behavior through organization *norms* and *values*. Many companies work deliberately to create a company culture that shapes their employees' attitude towards the value of work and the importance of the company's customers. These norms and values are often evident in the company's general business principles, its hiring and promotion policies, its compensation and benefits packages, its personal behavior standards, and so forth. Such control resembles the ideological kind of norm setting that is evident in, for instance, political parties, religious groups, or not-for-profit, activist organizations.

Five coordination mechanisms

The following *five coordination mechanisms*, as identified by Mintzberg (1979), help us understand how an organization coordinates and controls its various activities:

1. *Mutual adjustment:*
   Coordination occurs informally by direct communications between individuals. An example is the coordination in a team where everyone knows each other and each other's responsibilities well.
2. *Direct supervision:*
   Coordination occurs when an individual supervises other employees' work. A manager issues orders and checks that employees follow these orders. An example is the coordination by a military officer in charge of an operation.
3. *Standardization of work processes:*
   Coordination occurs when the work is described in detailed procedures and instructions. Sometimes such instructions are built into the technical equipment and tools used. Examples are the screen designs of webpages of computerized administrative systems.
4. *Standardization of output:*
   Coordination is achieved through clear specifications of the intended work results. Typical examples are the workstations on an assembly line in which the output of each station is standardized so that the process can continue at the next station.
5. *Standardization of knowledge and skills:*
   Coordination is achieved when the participants in a process have specific education or training that allows them to coordinate their work through well-defined terminology and the application of

standard operating procedures. An example is the cooperation of different medical specialists involved in a complex surgical operation at a hospital. Another example (of informal coordination) is the coordination of traffic in a city where all licensed drivers have a common understanding of traffic rules and use appropriate driving behavior learned in similar driving lessons.

In summary, we conclude that coordination and control in an organization do not always require a supervisor who closely monitors and checks the work. Rather, the opposite is true. Most daily activities at large companies are coordinated without the direct involvement of a supervisor. Much of the coordination is governed by the company's routines, rules, roles, norms, and values. To a great extent, these are influenced by the employees' education and experience.

3.2 Structure – more than an organization chart

If you ask employees to describe their company's organization, they will usually refer to the company's official *organization chart* (also called organogram). Although it may be drawn in various ways, the organization chart presents the official picture of how the organization is assumed to function and how the responsibility within the organization is distributed. A traditional organization chart consists of boxes that are linked with lines so that they form a *pyramid*. See Figure 3.1. Each box represents a person or a department (or unit) in the company with lines of authority drawn between them. Thus, the chart depicts the organization's *hierarchy* of superiors and subordinates. The chart answers the following questions: Who is the boss of whom and of which activities? Who has the authority to give orders to whom? And who should act as judge if two employees disagree?

The organization chart in Figure 3.1 is a typical example from smaller companies. In principle, however, the chart is generic and valid for most formal organizations – privately owned companies, public organizations, and not-for-profit organizations. The Board of Directors appoints the Managing Director (sometimes called Chief Executive Officer, CEO). He or she is the company's most senior executive and reports directly to the Board. Below the Managing Director/CEO, the organization is structured

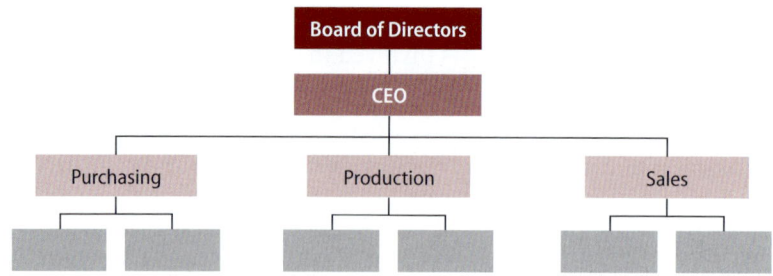

Figure 3.1 Classic organization chart.

into various vertical levels and into various horizontal departments with different specializations and responsibilities.

The formal and the informal organization

The organization chart presents the essential elements of the company's *formal organization structure*. The idea behind the formal organization structure – how the work is allocated and organized – is that it presents the most effective way of achieving the company's objectives. In addition, the organization chart is a representation of the company's various relationships as defined by job descriptions, instructions, and procedures.

Even if the departments shown in the organization chart appear independent of one another, it is a mistake to assume this is how the company operations actually function. An *informal organization structure* of personal contacts, friendships, and common interests among employees is at least as important (if not more important) as the formal organization structure. This informal organization usually plays an essential role in making the daily operations work smoothly and with high efficiency. If employees rigidly follow the formal structure, they usually accomplish less at a slower pace.

The formal and informal organization structures exist side by side. While the formal organization structure is impersonal and (in theory) independent of individuals, the informal organization structure consists entirely of individuals, their characteristics, and their personal relationships. The management challenge is to mesh the two structures so that they complement rather than oppose each other.

Positions, authority, and responsibility

The work of an organization is defined in accordance with how the various formal jobs and positions are located on the organization chart. Before creating a position in the organization structure, a company must identify the tasks as well as the authority related to that position. For example, does the individual in the position have the authority to sign contracts or to hire new employees? Typically, the authority of a position allows the employee to *delegate* some work tasks and decisions to his or her subordinates. Such authority may or may not be written into formal procedures.

However, even if it is possible to delegate authority to subordinates, it is not possible to delegate the responsibility. Anyone who delegates a task to someone still retains responsibility for its completion. Thus, it is always essential to define clearly which tasks have been delegated and which have not.

3.3 Various organizational forms

The organization structures of companies vary for several reasons (e.g., company size and age, industry sector, and business environment). For many years, researchers and practitioners tried to formulate general rules and principles for the ultimate organization structure. Although many of these principles are still valid, few researchers today accept the idea that there is one best way to organize (i.e., a particular organizational form is always preferred to another). Quite the opposite is true. Today there is a common understanding that an efficient organization structure can take many different forms.

The design of an organization structure can be examined according to three aspects: the division of labor; the integration of its activities; and the organizational control of employee behavior and performance.

Division of labor:

- horizontal specialization
- vertical specialization.

Integration of activities:

- vertical relationships: superior-subordinate
- horizontal relationships: associate-associate.

Organizational control of employee behavior/performance:

- accountability and performance.

Next we examine these five organization dimensions sequentially as we review the organization designs associated with each of them.

3.4 Division of labor

Horizontal specialization

A principal aim of organizing is to structure work tasks and teams into separate departments, units, sections, and so forth. This practice is called *horizontal specialization*. This division of labor depends primarily on the competences in the organization in relation to the work performed. In theory, we can identify four structures for grouping positions into departments and other organization units:

- function
- product
- customer
- geography.

The functional structure

The *functional structure* is one of the most common ways to structure the operations of industrial companies. The work is grouped according to functional expertise. The work tasks are divided in several main areas such as product development, manufacturing, and sales. Each area employs people with relevant expertise who develop the methods and procedures for their work. Through specialization, they create swift problem-solving routines while they develop their long-term competences. For example, by group-

Figure 3.2 The functional structure.

ing all experts on solid mechanics in the same department (vs. locating the experts separately in various departments) conditions are created that allow the company to depend less on each expert's individual knowledge. See Figure 3.2.

Of course, there may be disadvantages with the functional structure. Various teams and departments (with their experts) may tend towards isolation when they focus on their own tasks rather than on what is best for the company. The functional structure has a tendency to over-emphasize completion of functional tasks without taking the overall business into consideration. Also, a great many departments are inclined to engage with every problem/issue. This can create coordination problems for management.

The product structure

The *product structure* is another commonly used organization design. It is especially appropriate when the company produces goods and services that are significantly different from each other. In the pure form of this organization design, each product department takes responsibility for its own purchasing, operations, and sales. In this way, each department avoids many of the coordination problems that the functional structure experiences. Another advantage of the product structure is that it creates a holistic view of each product's requirements and characteristics. Furthermore, it may be easier to control operations if there is only one department responsible for each product. However, a disadvantage of the design is that every department requires its own specialists, which means that certain functions and tasks are duplicated throughout the company. Atlas Copco is an example of a global manufacturing company that uses the product structure. See Figure 3.3.

Figure 3.3 The Atlas Copco corporate structure. Source: Atlas Copco, 2014.

The customer structure

The *customer structure* is a way to group departments and teams by customers and customer categories. For example, the sales operations may be structured in one department for commercial customers, one department for non-commercial customers, and one department for export customers. The significant advantage of this organization design is that it minimizes the number of contact points between the company and its customers. Because specific departments manage different customer categories, the departments have good knowledge of their customers' needs and preferences. The sales departments of industrial companies often have sales representatives called *account managers* and senior sales representatives called *key account managers* who are assigned to develop and coordinate the sales activities for specific major customers. The disadvantage of this organization design is that it reduces the benefits of the functional organization from supporting long-term development of functional expertise.

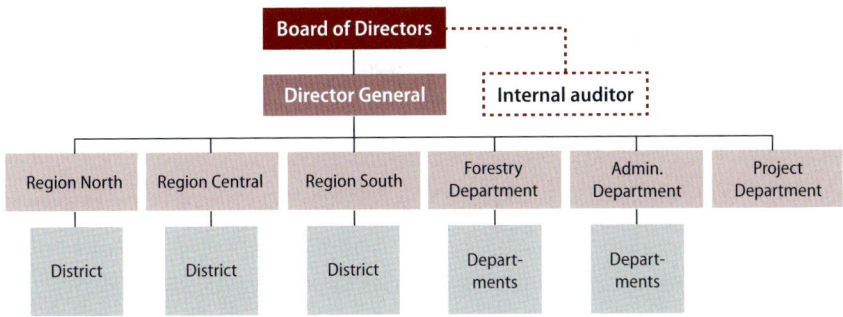

Figure 3.4 Swedish Forest Agency's geographic structure.
Source: Swedish Forest Agency (Skogsstyrelsen), 2014.

The geographic structure

The *geographic structure* is a way to group departments on the basis of the physical workplace locations. In many respects, this organization design is the simplest to understand and implement, especially when a company has operations at multiple locations in many countries. See Figure 3.4. The structure is often more effective if the employees at one location, who work closely on a daily basis, also report to the same manager. The advantage of this organization design is that it creates possibilities for developing local knowledge. The disadvantage is the same as with the customer structure – the difficulty of developing expertise in the various functional areas. However, many manufacturing companies that act in a global market have little choice. They are often obliged to set up local subsidiaries for product development and/or manufacturing in order to obtain contracts from the local customers.

Mixed organization structures

In reality, most large companies use a mixture of these four organization structures. Different principles are applied in different areas and at different hierarchical levels. Major industrial companies tend to structure their operations into business areas and subsidiaries by products, customers, or geographic locations. At the lowest organization level of the hierarchy, however, most industrial companies use the functional structure. Atlas Copco, for example, is not entirely structured by product. The Swedish Forest Agency is not entirely structured on a geographic basis. Instead these organizations have matrix structures, which are discussed below. See also Figure 3.9 (p. 65).

Vertical specialization

Most large organizations have a formal hierarchy based on responsibility and authority that ranks individuals and departments. In other words, the structure is based on a system of superiors who have more decision-making power than their subordinates. It is usual to talk about the different *decision levels* in which people with a superior rank in the hierarchy have more authority to make important decisions than the people below them. The situation in the box example (below) illustrates a typical hierarchical decision-making situation.

Hierarchy and span of control

Vertical specialization refers to the number of organization levels – that is, whether the hierarchy is flat (broad) or tall (narrow). The number of levels also determines the size of the *span of control*, which is a measure of the number of subordinate employees, or departments, that a manager coordinates. The optimal size of the span of control is one of the most debated issues in the history of organization theory. A classic organization principle, stated by the French industrialist Henri Fayol (1916), maintains that a manager should have no more than six subordinates. In a large organization, this would require an enormous number of hierarchical levels. More recent organization literature instead focuses on the advantages of the flat organization that broadens the span of control. Figure 3.5 summarizes the advantages of the flat and the tall span of control.

Hierarchies may in some sense seem old-fashioned and undemocratic. However, in many cases, hierarchical levels are a practical necessity, especially in organizations that must respond swiftly to urgent problems. When swift decisions are needed, not all employees can be involved in these decisions. Fire fighters, the police, and the military provide the purest examples of hierarchical organization designs where even clothing (the uniform insignias) makes decision-making authority visible. Moreover, if a superior in this kind of organization (e.g., a military officer) is absent, injured, or even deceased, it is a simple matter to determine who should assume the position of authority.

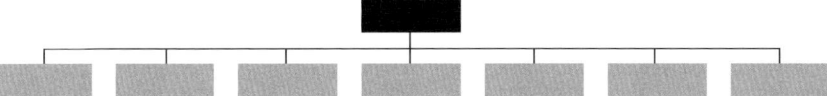

Flat (broad) span of control is suitable under the following conditions:
- A high degree of work standardization
- A high degree of similarity in work tasks
- A strong need for employees' independence and self-reliance, and
- A strong need to reduce disturbance in the vertical information flows in the hierarchy

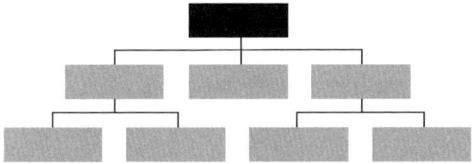

Tall (narrow) span of control is suitable under the following conditions:
- A strong need for direct management control of work tasks
- A strong need for reciprocal adaptation to manage work tasks that are tightly linked to each other
- Employees have a strong need to contact managers for help and advice, and
- Managers have many other work tasks besides their managerial responsibilities

Figure 3.5 Advantages of different spans of control.

EXAMPLE

At the company, Constellation AB, none of the employees has the authority to purchase materials or equipment or to register for mobile telephone subscriptions. If employees need work gloves or new computers, for example, they must check with their managers before making the purchases. The managers, in turn, can only make purchase decisions for items below SEK 20,000 or less. For larger amounts (e.g., purchases of new machines or expensive training programs), the managers must check with the production manager. In turn, the production manager is limited to making purchase decisions of up to a couple hundred thousand SEK. For more costly purchases, only the CEO, in consultation with the Board members, has the decision authority.

The exception principle

A hierarchical organization works very well for coordination by direct supervision. See Section 3.1. While employees are responsible for completing their work tasks within specified frameworks, their managers create those frameworks, check that employees follow the rules, and control that work is completed as planned. The *chain of command* in principle should be consistent with the formal hierarchy. Thus, a top manager doesn't issue orders directly to employees at the lowest hierarchical levels; rather, the manager issues orders that flow down the chain of command in the hierarchy to the appropriate levels.

Employees in a functioning hierarchy perform many tasks without managerial supervision. As long as they perform these tasks according to established routines and standards, their managers do not need to be involved. However, when employees have tasks that fall outside the established routines, they must ask for advice from their supervisory managers. In this way, employees handle routine work tasks while their manager focuses on other problems and difficulties. Scholars of organization theory have used the term *the exception principle* to describe this situation. The principle proposes that a manager at a higher level will deal with the most critical problems and the most difficult and unusual work tasks (i.e., the exceptions) while the subordinate personnel will focus on the ongoing operations and minor problems. We see this principle in many everyday situations. As an example, store clerks forward customer complaints to their store managers. We also find the principle in national juridical systems when legal cases escalate from local courts to appeals courts, and from appeals courts to supreme courts.

3.5 Integration of operations

After activities have been structured in an appropriate manner, they must be integrated so that the specialized departments (units, subsidiaries, etc.) work toward the common objectives.

Vertical relationships: superior-subordinate

Naturally, there must be established relationships between the various organization levels. Vertical relationships deal with the relationships (especially via communications) between superiors and subordinates, and between subordinates and other subordinates.

The line organization structure

Perhaps the most well-known organization structure is the *line organization* (cf. line of command). Its basic principle is the scalar principle. This principle states that authority is distributed top down so that every employee has one, and only one, direct senior manager, and managers have authority only over their closest subordinates. See Figure 3.6. The line organization structure, which is the traditional way to draw the organization chart, is inspired by the military services. The advantage of the line organization structure is its simplicity and clarity, which makes it quite suitable for direct supervision. A company that consistently uses line organization typically has a very rigid and formal organization structure.

In practice, *staff positions* are used to adapt or modify the line organization structure. Staff positions, or units, are outside the formal line of command. People in these positions lack formal decision authority but can offer advice to decision-makers as well as to the departments in the line of

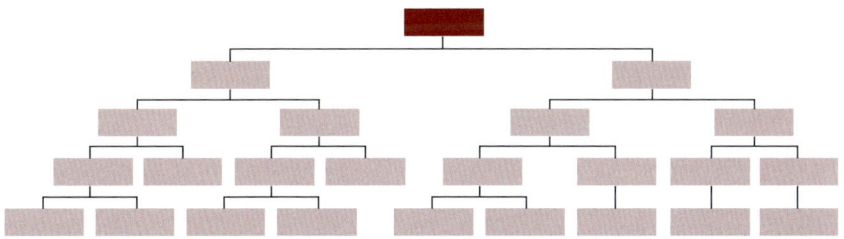

Figure 3.6 The line organization structure: The principle of unity of command in which each subordinate has only one superior.

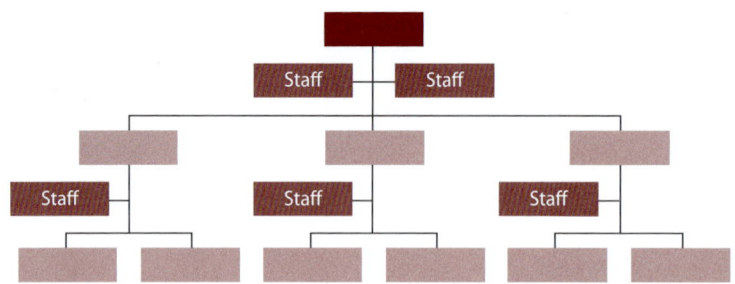

Figure 3.7 The line-staff organization.

command. This concept – the *line-staff organization* – also has its origin in the military services. See Figure 3.7.

Usually staff positions exist at several levels in the organization. Therefore, staff units can be of different sizes and have different responsibilities and functions. The organization chart in Figure 3.7 depicts how a small company is organized. The Managing Director is at the top level with two staff members (e.g., a secretary and a financial assistant). The three Department Managers are at the next level with their staff assistants. Note that these staff assistants are outside the direct line of command. The same general organization structure is found at large, global companies in which each business area has its own staff.

There are two categories of staff units: (1) support and expert units that support management and provide expertise to the operational units; (2) general service units that provide various services to the operational units.

The unity of command and staff units

In practice, the classic principle of unity of command is disappearing at most companies. From experience, the pure line organization structure has proven to be unsuitable when a business grows and becomes increasingly complex. Actually, the introduction of staff units even violates the principle of unity of command.

A small or medium-size company can rarely afford to employ experts at every level (e.g., experts in sustainability, healthcare, or legal issues). Instead, members of the staff units attached to top management typically take the role of such experts. In theory, a company's legal counselors (attorneys), for example, can only offer advice to the purchasing and sales departments; they cannot make managerial decisions. However, this is rarely the case in practice. Because of their specialized, legal expertise, a legal counselor can often exert a strong influence on purchase and sales contracts. It is doubtful that many companies would enter into large contracts against the advice of their legal counselor.

Thus, line managers seldom have the clear-cut authority that the organization chart indicates. Even when they have full formal responsibility, their autonomy is often limited in practice by different directives, instructions, and company procedures as well as by established norms and values among the employees.

Functional work supervision

Frederick W. Taylor, a pioneer of industrial management theory and practice in the early 1900s, offered another view of management – *functional supervision* (not to be confused with the functional organization structure described above). An essential element of Taylorism is that the division of work into very small tasks makes manufacturing more efficient. The theory also proposes that there are advantages from dividing and specializing the

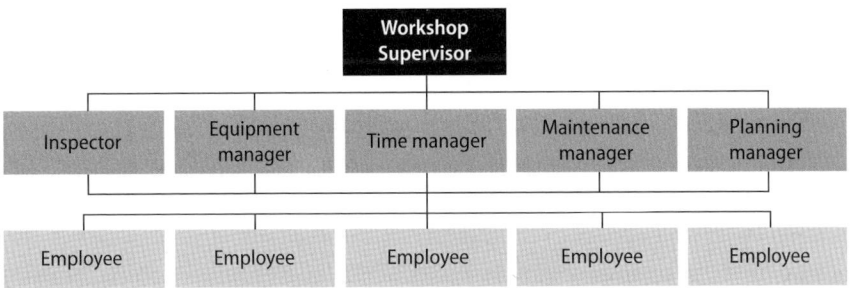

Figure 3.8 Functional supervision.

work of managers as well. Depending on the specific task at hand, different functional supervisors should issue the orders to employees. In the example (above), each of the three managers has a different responsibility. See also Figure 3.8.

The principles behind functional supervision may seem somewhat strange. Early on, advocates of unity of command challenged these principles. From the employees' perspective, however, the difference between functional supervision and line-staff organization is not particularly great. The difference between orders from a manager and advice from an expert (e.g., a design engineer or a production engineer) is often rather subtle.

EXAMPLE

Functional supervision: Supervisor A, the inspector, is responsible for ensuring that work instructions are followed, that the work meets quality standards, and that the work is carefully and precisely performed. Supervisor B, the production manager, shows the employees how to operate the machinery and how to perform the operations. Supervisor C, the time manager, controls that the machinery works at the proper speed so that throughput time is minimized. In addition, other managers, such as the maintenance manager and the production-planning manager, direct and assist the workers.

Horizontal relationships: associate-associate

The simplicity of the organization chart easily creates over-confidence in the ability of leaders to manage an organization using a rigid and hierarchical format in which all communication follows the line of command. However, managing an organization is not that simple. In reality, employees from different departments coordinate many of a company's daily activities through both formal and informal communication channels (the coordination described earlier in the chapter as "mutual adjustment"). Often, the company tries to enhance these types of horizontal relationships. Formal horizontal relationships can be designed in several ways. Galbraith (1973) offers the following examples:

- liaison roles
- standing committees
- temporary teams and projects
- matrix structures
- networks.

Liaison roles

A simple way to improve communications between two departments, such as product development and manufacturing, is to appoint a design engineer as the contact person for manufacturing issues. In this *liaison role,* the design engineer can simplify and streamline the inter-departmental communications. Employees in the manufacturing operations know they should contact this engineer who also understands their problems. If the two departments are located some distance apart, the engineer may even have a workstation at the manufacturing plant where he or she acts as the go-between "ambassador".

Standing committees

It is also quite common to create horizontal relationships that span organization boundaries using inter-departmental *standing committees.* These committees are composed of people from different departments who meet regularly to discuss issues of common interest. Some examples are committees for purchasing decisions, changes in product designs, and company-wide, quality control measures. Although a standing committee may not have any formal decision authority, it may still be very important for the company. When actors from different departments meet at regular intervals, many inter-departmental issues can be resolved through agreement without the involvement of supervisory managers (i.e., by mutual adjustment). Moreover, the exchange of information and knowledge among the committee members can counteract the tendency towards an inward oriented, narrow focus that easily emerges in highly specialized organizational units.

Temporary teams and projects

A third way to create horizontal relationships is to use *temporary teams.* These teams are formed to solve problems that are unusual, quite difficult, and require input from many departments. An example is a temporary team established to plan office space design and the allocation of workstations when a company moves to a new location. Or a company may form a temporary team to investigate an unexplored market or to study the company's need to invest in a new accounting system. It is quite common to create such teams for investigations of various kinds.

A special kind of temporary team is the *project organization.* Originally, such temporary project organizations were used for extremely large or extraordinary projects (e.g., development of military jet fighter planes or construction of nuclear power plants). However, today project teams are used to coordinate many types of undertakings, such as product launches, capital investments, or changes in manufacturing machinery. Although there are no major differences between projects and other temporary teams, project organization is typically used for particularly important and long-lasting, formal undertakings. A project manager and a team of specialists are needed although they are not always permanent company employees. The project team is tailored as far as possible to match the specific task at hand. Major projects may involve many employees on a full-time basis for a long time. Most projects, however, engage many of their team members on a part-time basis.

Matrix structures

A fourth way to enhance horizontal relationships is the *matrix structure.* When applying this organization design, a company deliberately moves away from the line organization's basic idea of unity of command. Instead, the company superimposes cross-functional coordination lines on top of the functional lines of command for the departments. This kind of organization is common in many organizations. In large engineering companies, for example, it is customary that project managers have responsibilities that span the compartmentalized, functional departments. Furthermore, in

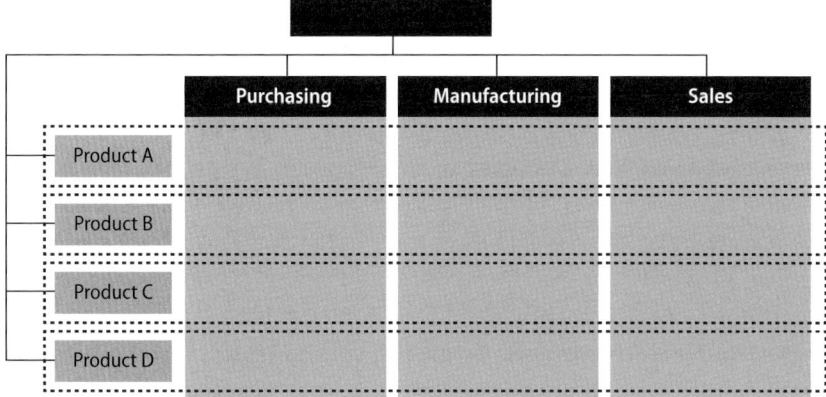

Figure 3.9 The matrix organization structure.

global corporations it is common that the traditional functional structure is combined with a geographic structure for different markets. See Figure 3.9. Thus, matrix structures involve at least two organizational dimensions, and in some cases many more. Depending on the nature of the issues at hand, the employees work with, and are coordinated by, different managers (cf. Taylor's functional supervision).

The matrix organization structure tries to combine the advantages of other organization designs. However, this is not easy. In most organizations there are latent tensions between different departments such as, for example, tensions related to product responsibility vs. functional responsibility. These tensions have a tendency to surface in matrix organizations.

Formalized or not, most major industrial companies use the matrix organization structure in some way. For example, employees from different departments work together on various projects under the direction of various project managers. Thus, in order to be efficient, a matrix organization structure requires shared responsibility and close cooperation between, for example, the cross-functional project managers and the functional department managers who "lend" their employees to the different projects.

Networks

Networks are the fifth way to enhance horizontal relationships in a company as well as relationships with actors outside the company. Networks, or communities, are composed of actors with a common interest and identity. These relationships may be as strong as, or stronger than, the formalized relationships within a company. Many professions, such as physicians, attorneys, and accountants, have powerful professional norms and often very formal rules that control their work procedures and behavior in various situations. In addition, many company employees are members of professional networks established by contacts formed in college, executive education and vocational training, trade unions, or other professional interest organizations. Many companies encourage employee participation in networking and even initiate the creation of networks for different purposes. One reason might be to enhance the professional development of the company's project managers. Another reason might be to support the professional careers of the company's female managers.

3.6 Accountability and performance measures

However, specialization and coordination alone do not determine employees' ways of working or their attitudes/behaviors. Performance measurement is another important factor in this regard. Many companies actively apply the classic motto "What gets measured gets done". Thus, with the use of many financial performance measures, companies try to motivate and control results. This is an old idea that is often attributed to the American business executive, Alfred P. Sloan, who introduced the concept of profit centers in 1920 at the automotive company, General Motors.

In the following, we describe four financial/organizational concepts commonly used in discussions on performance management:

- revenue center
- cost center
- profit center
- profitability center.

A *revenue center* has responsibility for the sales of the company's products (goods and services) and the associated resources behind these sales. The natural revenue center is the sales department. Although its activities are conducted on behalf of the entire company, the center is responsible only for its own costs. The center has sales targets, such as sales volume per month or week. Employees in the department may have individual sales goals. Meeting these goals may result in some form of performance bonus.

A *cost center* has responsibility for its own costs, but it generates no revenue. Therefore, the center does not make a profit. For example, a manufacturing department is often a cost center that can only influence its costs by producing the products more efficiently (i.e., by producing faster and cheaper or by reducing waste). A cost center usually has a monthly or yearly cost budget. Tight budgetary control is maintained by specifications on manufacturing cost per product. See Chapter 4.

A *profit center* has both costs and revenue responsibility. The most important feature of the center is, however, not the amount of its costs and revenues but rather the difference between them (profit = revenues – costs). This feature gives the center a broader framework for its activities than if the costs and revenues were determined in advance. The center can increase its profit in two ways: increasing revenues or reducing costs.

A *profitability center* tries to maintain a certain level of profitability. This means the center tries to achieve a certain return on the capital employed in the operations (such as machinery, etc.). At large companies, *divisions* or *business units* are usually designated as profitability centers.

In principle, business units (or divisions) are "companies within the company". Top management of a company with a business unit structure usually manages each business unit on a strategic level with profitability objectives in focus. Management does not get involved with the details of the business units' daily operations. Therefore, a business unit often has complete responsibility for the business and products within its area.

If a company wants still greater decentralization of authority, it can *incorporate* business units/divisions as legally independent *subsidiaries*, while still retaining ownership of them. At this point, the corporation forms a *group* (the Swedish word for "group" is *koncern,* which derives from the German *konzern)* with one *parent company* that has control because of its majority ownership of one or several subsidiaries.

3.7 Conflicts of interest and objectives – the organization's largest problem?

We often hear executives explaining "the organization's objective is …", or "our objective is …", or something similar. Perhaps the business concept is the most explicit expression of an organization objective. However, few organizations have only one objective. The members of an organization often have individual interests that may not always coincide with those of the organization. For example, parents may want to relax on the weekends with their children, but the company may prefer that the parents come to work.

Different organization objectives often conflict. For example, one objective of a city council may be to increase efficiency by decreasing turnaround solution times for its residents' various social problems. However, because the city council is required (by law) to treat all residents equally and cannot treat some residents more favorably than others, more detailed investigations and longer turnaround times may be necessary. In addition, different parts of an organization may have different objectives and interests. A typical example of such a *conflict* of interests is when a company's marketing manager wants many versions of a particular product to meet different customer demands, while the manufacturing manager, for reasons of costs and efficiency, wants as few versions as possible.

Setting organization objectives, then, involves negotiations among the various stakeholders, within and outside the organization. Nor are organization objectives fixed and permanent. They are always under review and revision. Thus, coordinating and controlling organizational behavior are very complex activities. These activities are far more complicated than they may have appeared in this chapter.

Finally, organizing is not merely a matter of designing and redesigning the organization chart. Organizations are groups of people, with feelings, who work together to accomplish something!

CHAPTER 4

Product costing – estimation, determination, and allocation

4.1 The basic concepts of accounting and costing

There are many reasons to analyze a company's financial situation. One of the more common reasons is to obtain information for important decisions related to the company's future. Should we develop a new product? Will we make money on a particular service? How much could we lose in a given market on this product or service? Should we shut down a department?

Making such financial decisions means using established and standardized methods for obtaining data and for making calculations. Obtaining this data often requires making various financial estimates and calculations.

In this chapter, we present various concepts and methods used to make such estimates and calculations.

Costs, revenues, and other concepts

There are three significant pairs of concepts that are used to describe a company's money inflow and money outflow: *revenues* and *costs*, *income* and *expenditures*, and *receipts (cash in)* and *payments (cash out)*. These terms are defined as follows:

- *Revenues* are the value received for a product or service delivered in a certain time period.
- *Costs* (expenses) are the value of resources consumed in a certain time period.
- *Income* occurs in connection with a company's sending of an invoice.
- *Expenditures* occur in connection with a company's receipt of an invoice.

- *Receipts (cash in)* occur when a company receives cash payments.
- *Payments (cash out)* occur when a company makes cash payments.

When a company buys a product or a service from a supplier, the company will receive an invoice for the delivery. By definition, the *expenditure* occurs when this invoice is recorded in the company's bookkeeping system. At the same time, the *income* occurs for the supplier when the supplier records the amount of the invoice. The invoice states the payment's due date. On that date, at the latest, the purchasing company makes a *payment* (cash out), and the supplier records a *receipt* (cash in). In short, these dualities help us follow the sequential stages of business transactions.

Companies record these transactions – income, expenditures, receipts (cash in), and payments (cash out) – on a continual, day-to-day basis. See Chapter 6. The two concepts of revenues and costs, however, relate to a specified time period, usually a year although other time frames are also used (e.g., months, quarters, etc.)

Consider the following example: A company in a certain year purchases raw materials for a manufacturing process from a supplier but does not immediately put these materials into production. The company records the expenditure when the invoice for the materials is received and records a payment when the invoice is paid. However, the company has not yet incurred a cost (given the definition of costs) because the materials – a resource – have not been consumed. Thus, the expenditure is recorded in the current year, but the *cost* is allocated (deferred) to a future year when the resource is consumed. The cost is *allocated to the relevant time period*. The expenditure was incurred and recorded, and may have even been paid, in the current year, but the cost will not be recognized until the resource is consumed.

There is a similar situation with revenues and receipts. Receipts (cash in) may be received before they are earned, in which case the *revenue* is allocated (deferred) to a future year. These examples illustrate the concept of allocation in accounting for companies' revenues and costs and their income and expenditures.

When did the cost occur? According to our definition, a cost is the value of resources consumed in a certain time period; in other words, when were the raw materials in our example consumed? As already observed, the cost did not occur when the raw materials were ordered, delivered, or paid for. In fact, these materials continue to be resources until they eventually leave the company, often as part of a delivered product. In accounting terms, no costs exist until the resources are used in the company's production of a product that has been sold and delivered to a customer. In accounting, you sometimes use the term *cost of goods sold*, which is an accounting term for the cost of the materials used when the company sells the product that includes these consumed materials.

Many difficulties, however, arise in estimating and calculating the cost of goods sold. For example, what quantity of the materials was actually used for a specific product? How many hours were spent working on a specific product? How do we determine those amounts? Can we trust the reported data that we use for our calculations? Or is further investigation required? These are examples of important questions that must be asked and answered when making such estimates and calculations.

Fixed costs and variable costs

A company's *profit* is the difference between its *revenues* and its *costs* (expenses). A manufacturer's revenues derive primarily from the sale of its manufactured products. Thus, revenues are proportionate to sales – the more products sold, the greater the revenues.

However, a company's costs have other relationships. While some costs, like revenues, vary with volume, others do not. Therefore, it is necessary to differentiate between *fixed costs* and *variable costs*.

Variable costs relate to some *activity volume* – for example, to the number of products produced, to turnover, or to some other measure of a company's output or input. As such activities increase, so do variable costs. Fixed costs, however, are independent of the activity volume, at least within a certain range. Variable costs increase with the increase in activity volume. Variable costs may be linear, progressive (increasing), or regressive (declining). Fixed costs are constant at a certain level of activity although they may jump to a different (typically higher) level outside the relevant range. Such fixed costs, called *step costs*, then are fixed in a new relevant range where they are again independent of activity volume. For example, a new relevant range for fixed costs is set when business activity increases to the point that the manufacturing plant must be expanded. See Figure 4.1.

Fixed costs

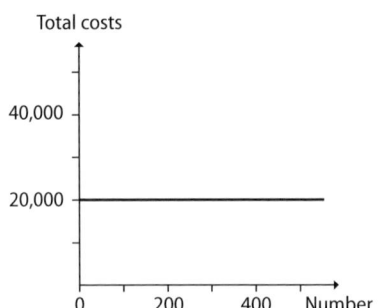

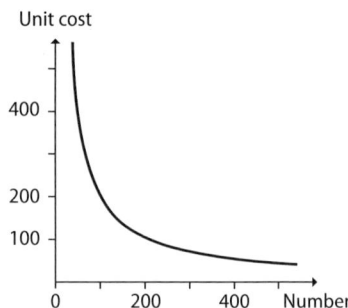

Step costs

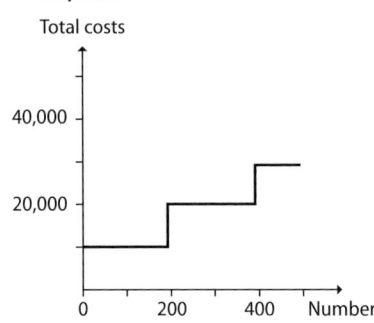

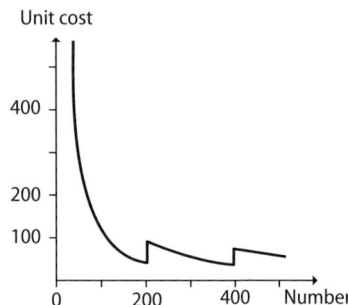

Linear variable costs

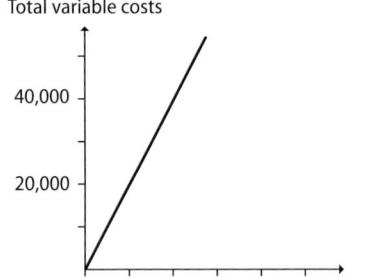

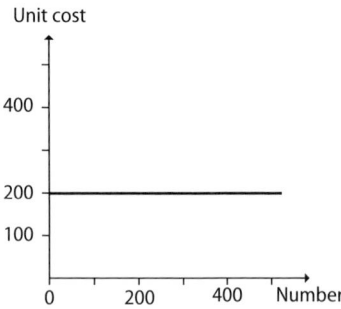

Progressive variable costs

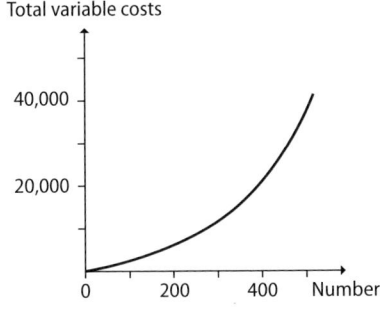 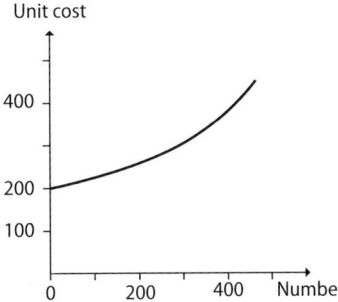

Regressive variable costs

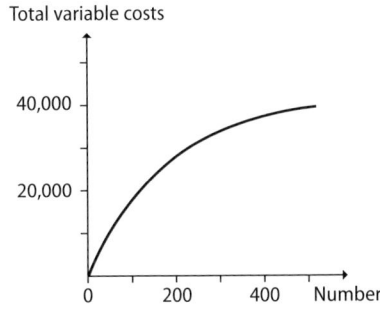 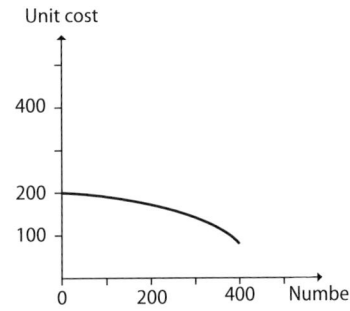

Figure 4.1 Fixed and variable costs.

To illustrate the relationships between total revenues and total costs (fixed and variable) we can draw the *break-even chart*. Figure 4.2 illustrates a case in which the total revenues are proportional to the activity volume, the total variable costs are proportional to the activity volume, and the total fixed costs are independent of the activity volume. A company that has both fixed costs and variable costs is said to *break even* when total revenues exactly equal total costs. In the break-even chart, total variable costs and total revenues appear as linearly dependent on volume.

Break-even chart formulas

Total costs = Fixed costs + Variable costs/unit · Volume

Total revenues = Price/unit · Volume

Break-even point or Critical volume: Total revenues = Total costs

Safety margin (volume) = Real volume – Critical volume

$$\text{Safety margin (\%)} = \frac{\text{Real volume} - \text{Critical volume}}{\text{Real volume}}$$

This is, of course, a simplification because, in reality, a large increase in sales volume often affects the revenue per unit, producing a regressive revenue curve. Few companies sell their total annual production at the same time to the same customer. Therefore, to set prices or to evaluate whether certain orders should be accepted or not, a company must first calculate the cost per product unit.

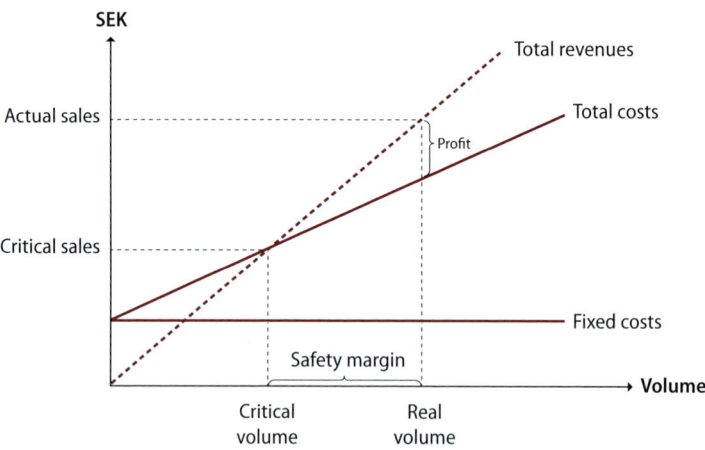

Figure 4.2 Break-even chart. Total revenues exactly equal total costs at the critical volume (break-even point) (based on Lantz et al., 2014).

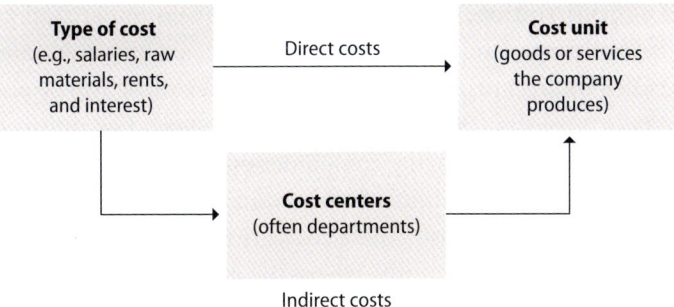

Figure 4.3 The cost allocation principles.

Direct costs and indirect costs

Costs can also be categorized according to their type: for example, salaries, raw materials, rents, and interests. See Figure 4.3.

In referring to *direct costs* and *indirect costs (overhead)*, we link these costs to various *cost units*. By cost units, we refer to the goods and services that the company produces (i.e., the revenue-producing items that are intended to cover the company's costs). Products or projects are common cost units at many companies.

Direct costs are relatively easy to trace to the different products. For example, the costs of raw materials, various production components, and labor for production employees, etc. can be traced directly to the manufacturing processes and to each product.

Direct costs

Costs that can be traced directly to specific cost units.

Indirect costs (overhead)

Costs from a department or function (a cost center) that are indirectly allocated to cost units.

Costs for administration, sales, and product development staff and for machinery, buildings, and top management, etc. are more elusive. Such costs are called *indirect costs* or *overhead costs*. They must also be allocated to products or projects (the cost units) in some way. For *indirect costs*, traditionally most manufacturing companies establish cost categories such as *materials indirect costs, manufacturing indirect costs, sales indirect costs,* and *administrative indirect costs*. These costs are then assigned to *cost centers* (i.e., the various organization units where the costs were incurred). When this is completed, the costs of the cost centers are further distributed to the various cost units (i.e., the produced goods and services or projects), typically by the use of standardized overhead charges (see the examples below).

Common costs and specific costs

Based on how costs behave, they can be differentiated as either *specific costs* or *common costs* (sometimes called "relevant costs" and "irrelevant costs"). Costs that are incurred as the result of specific actions are *specific costs*. Examples are the costs associated with beginning production of a new product or with processing an order for a particular product. Costs that are *not* incurred as the result of specific actions are called *common costs*. Examples are machinery costs that are unaffected by the decision to manufacture a new product or to process an order.

Thus, the situation determines whether a cost is a common cost or a specific cost. Moreover, in some situations, a cost may be classified as a common cost while in other situations the same cost may be classified as a specific cost.

The term *specific revenue* is used to refer to revenues affected by a specific action. In product costing, this refers to a product's sales price. After all, the sales price is the additional revenue from the sale of the product.

Cost unit

The objects (product, project, department, machine hour, etc.) for which a cost calculation is made.

Specific costs (relevant costs)

The costs that are directly caused by an action or decision in focus of the cost calculation.

Common costs (irrelevant costs)

The costs that are unaffected by the action or the decision in focus of the cost calculation.

A comparison of cost concepts

The division of costs according to these different concepts has different purposes. Each concept pair can be used to divide all costs. However, because the pairs overlap, every cost may be categorized according to each of the three basic principles. See Figure 4.4.

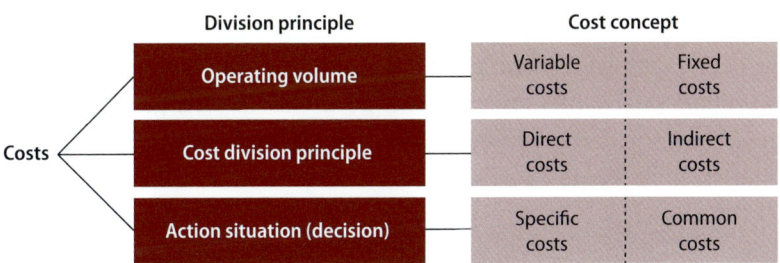

Figure 4.4 Cost divisions by three different principles.

4.2 Cost determination and product costing

A company needs to know in advance how much it will cost to produce a product or to offer a service. Such information is used to set prices or to calculate if a certain product or service will be profitable. Furthermore, this information is useful for decisions about production methods or in make-or-buy decisions (i.e., which components to make in-house and which components to buy from the suppliers). These are essential decisions that companies must make even before production begins. This is the background for what is often called *product costing*.

In this section we introduce four methods of product costing:

- Contribution costing
- Full costing
- Step costing
- Activity-based costing.

4.3 Contribution costing

In many companies it is often not necessary to allocate all costs to the cost units (i.e., goods or services) when calculating the cost to produce and deliver them.

A common method of calculating the product cost without allocating all costs is *contribution costing*. In this method, only the *specific costs* are allocated to the cost units. Thereafter, each cost unit's *contribution margin* (i.e., the coverage of the common costs) can be calculated. Specific costs are those costs that the company incurs because it produces a particular object (a product, an order, a project, etc.). These are costs that would not have been incurred had the company not produced that particular object. By comparison, *common costs* are the costs that the company would incur regardless of whether or not it produced the object. Examples are office costs, administrative salaries, and so on.

Contribution margin

Contribution margin (CM) for an object = Specific revenue for the object – Specific costs for the object.

The *total contribution margin*, the sum of the contribution margins from all cost units, should cover total common costs, and, hopefully, provide a profit. Contribution costing is used mainly for existing resources in situations where assortment, demand, and capacity remain constant in the short term. As an example, business hotels often charge lower rates on weekends. Most hotel costs are common costs (i.e., costs that the hotels will incur whether their rooms are rented or not). Therefore, the revenue from renting a room at the lower weekend rate may cover the room's specific costs (e.g., cleaning) and will contribute towards covering the hotel's common costs.

Contribution costing is commonly used in, for example, consultancy companies when determining if client assignments should be accepted at a certain price. The following example illustrates how contribution costing works.

EXAMPLE

The consulting firm, LexCom, provides systems and programing advice, among other things, to companies that want to adapt their systems for electronic sales. The following information is available for one period. Note: Project 3 has a negative contribution margin (CM).

	Project 1	Project 2	Project 3	Project 4	Total
Specific revenue (SEK)	800,000	1,200,000	1,500,000	500,000	4,000,000
Specific costs (SEK)	−350,000	−600,000	−1,550,000	−250,000	−2,750,000
Contribution margin (SEK)	450,000	600,000	−50,000	250,000	1,250,000
Common costs (SEK)					−800,000
Period's profit (SEK)					450,000

The LexCom example shows the contribution margins for the four projects. The analysis is important because it reveals at least two things. First, the company can see each project's contribution margin. Second, the analysis reveals that common costs must be subtracted from the total of the four projects' contribution margins to calculate company profit. For consulting firms, in which the basis for overhead charges is the same for all projects (the consultants' work hours), contribution costing is often used because it simplifies the calculation when preparing project proposals for new clients.

4.4 Full costing

It is obvious that contribution costing has a big disadvantage. When using this costing method, only the specific costs are allocated to the cost units. This means that it is impossible to know if *all* costs are actually covered at a certain price for the goods and services that are produced.

In *full costing*, a company calculates all costs related to a product until it is delivered and payment is received. The company that manufactures only one product illustrates the simplest form of full costing. This calculation is often called *process costing* and means that all costs of operations are divided by the production volume. Since the company only manufactures a single product it needs to carry the total costs. In the Trebla AB example (see below), we can rather easily conduct an investigation and acquire relevant data (e.g., from employee interviews) for this type of costing. The situation becomes more complex when a company manufactures several products. Such an analysis is not possible at a company with multiple products, in the hundreds or even thousands. In that situation, some form of a systematic cost allocation method is required.

In full costing, direct costs and indirect costs are handled differently. Direct costs are not problematic because the company's cost accounting system traces these costs to the cost units. The indirect costs (overhead), however, must be allocated among the cost units.

> **Full costing for a product**
>
> In full costing, the cost of a product is the sum of all costs until the product is delivered and payment is received.

EXAMPLE

The company, Trebla AB, manufactures only one product: a chair. The company's total costs to manufacture 40,000 chairs in the previous year were SEK 6,000,000.

How much does it cost to manufacture and deliver a chair?

The full cost for a chair $= \dfrac{\text{SEK } 6{,}000{,}000}{40{,}000 \text{ chairs}} = $ SEK 150 /chair

Recently the company received an order for a variation on its standard chair that costs SEK 150 to manufacture. The original chair was an easy chair, but the new order is for an office chair. The company wants to calculate the cost to manufacture 20,000 Easy Chairs and 20,000 Office Chairs.

It is no simple matter to work out the costs to manufacture and deliver two different products. Although many company resources can be used for both chair types, they are used in different amounts. Among these resources are warehouse personnel wages, the cost of warehouse equipment, managerial, sales, and administrative salaries, and utilities costs. The only cost that is easily traced to the two chair types is the cost of the direct materials.

Trebla then conducted an analysis of these costs. Employees were asked to determine how much time they spent on the two production activities. Employees used their experience with Easy Chairs (as far as time spent and materials used) for their calculations for Office Chairs. Their calculations differed widely, so Trebla averaged these amounts, as shown below.

ch = chair; hr = hour

Data from Trebla personnel. All figures are shown in SEK.

	Easy	Office	Total cost
Materials cost	SEK 40/ch	SEK 60/ch	2,000,000
Warehouse, etc. (% of time)	40 %	60 %	400,000
Time worked	0.20 hr/ch	0.28 hr/ch	2,000,000
Machine cost, etc. (% of time)	45 %	55 %	1,200,000
Sales, etc. (% of time)	60 %	40 %	400,000

Based on these data, Trebla made the following calculations for the manufacture of 20,000 Easy Chairs and 20,000 Office Chairs.

	Easy	Office	
Direct materials cost	800,000	1,200 000	
Indirect costs (overhead) for materials handling	160,000	240,000	40 % and. 60 % of 400,000
Direct labor costs	833,000	1,167,000	0.20/0.48 and 0.28/0.48 of 2,000,000
Indirect costs (overhead) for manufacturing	540,000	660,000	45 % and 55 % of 1,200,000
Indirect costs (overhead) for sales and administration	240,000	160,000	60 % and 40 % of 400,000
Full cost:	2,573,000	3,427,000	
Full cost/chair	128.65	171.35	

Absorption costing is a systematic way to allocate costs to cost units. Direct
costs are traced directly to products, but the indirect costs (overhead) are
allocated at flat rates to the cost units.

Figure 4.5 illustrates the most common way manufacturing companies
calculate the cost of producing a product using absorption costing according
to the *full costing method*. The calculation involves the following costs:

- *direct materials costs (dM)*:
  costs of materials used in production.
- *indirect material costs (materials overhead; MO)*:
  costs associated with handling materials, such as storage costs and
  warehouse personnel.
- *direct labor costs (dL)*:
  wages of the factory workers for manufacturing the product.
- *indirect manufacturing costs (manufacturing overhead; MgO)*:
  costs associated with manufacturing, such as machine maintenance
  and production planning.

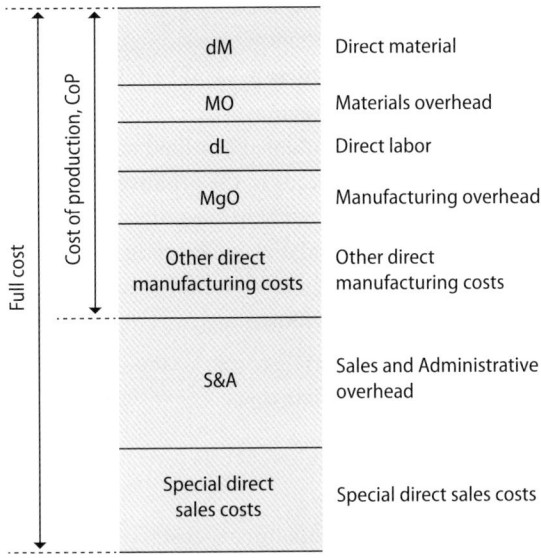

Figure 4.5 The components of the full costing method.

- *other direct manufacturing costs:*
  for instance patents and licensing fees.
- *indirect sales costs (sales overhead; SO):*
  costs associated with marketing, sales and advertising.
- *indirect administrative costs (administrative overhead; AO):*
  costs associated with management, financial issues, etc. SO and
  AO are often combined and then called *Sales and Administrative
  overhead (S&A).*
- *special direct sales costs:*
  (e.g., commissions).

In order to make cost allocations as fair as possible, it is important to determine the basis for each *cost allocation,* which means finding a link between
each cost and each cost unit. The aim is to match, as closely as possible,
the consumption of resources to each of the products. In many companies,
equipment cost is a major cost of production. If certain equipment is used
exclusively for certain products, the equipment cost can be considered a

EXAMPLE

If Trebla manufactured many products, it would not be possible to make these cost allocations in the way previously described. Instead, it would be necessary to determine the appropriate allocation bases and methods for assigning the indirect costs (overhead) to the products. The costs of storage and handling are probably proportional to the number of products manufactured. However, how much material does each product require? The company can calculate a relationship between total indirect materials cost (overhead) and total direct materials cost, as follows:

$$\text{MO charge} = \frac{\text{Total indirect materials costs}}{\text{Total direct materials costs}} = \frac{400{,}000}{2{,}000{,}000} = 20\ \%$$

In the same way, manufacturing indirect costs (overhead) can be expressed as a percentage of total direct labor costs.

$$\text{MgO charge} = \frac{\text{Total indirect manufacturing costs}}{\text{Total direct labor costs}} = \frac{1{,}200{,}000}{2{,}000{,}000} = 60\ \%$$

Trebla must also allocate sales and administrative costs to the products. Similarly, these costs can be expressed as a percentage of Total Cost of Production (CoP).

$$\text{S\&A charge} = \frac{\text{Total indirect Sales \& Administrative costs}}{\text{Total Cost of Production}} = \frac{400{,}000}{5{,}600{,}000} = 7.2\ \%$$

With this information, we can now apply full costing to Trebla's products. All figures are shown in SEK.

	Easy	Office	
Direct materials cost (dM)	800,000	1,200 000	
Indirect material costs (MO)	160,000	240,000	20 % of dM
Direct labor costs (dL)	833,000	1,167,000	
Indirect manufacturing costs (MgO)	500,000	700,000	60 % of Dir. Man. costs
Cost of Production (CoP)	2,293,000	3,307,000	
Indirect sales and administrative costs (S&A)	165,000	238,000	7,2 % of CoP
Full cost	2,458,000	3,545,000	
Full cost/chair	122.90	177.25	

direct manufacturing cost. On the other hand, the wages of the equipment operator (who may work on several machines) is an *indirect manufacturing cost* (overhead) that must be allocated based on time worked per machine.

When we compare the costs of the two chairs on the basis of this calculation with the analysis made in the first example, we notice that they differ. A fundamental weakness is apparent in the allocation method: a standardized method will produce an approximate result. Yet, at the same time, that is the method's strength. We can use the method to make allocations for products when the products are so numerous that it is impossible to make detailed analyses.

The *absorption method for full cost allocation* is based on the fundamental assumption that the relationships between the direct and indirect costs for the company as a whole are also valid for each of the individual products. For Trebla (see the example below), the same *overhead charges* (percentage rates) and the same *bases for overhead charge* are used for all products. Since these costs are indirect, they cannot be easily associated with specific cost units. Instead, they must first be attributed to different cost centers. If it is possible to determine a basis for the overhead charge for one specific product, it means that these costs actually are direct costs related to this product, and consequently no indirect cost allocation is required.

Using absorption costing

When a company uses the absorption costing method and makes *indirect cost allocations*, the work requires four steps:

1. definition of bases for the overhead charge
2. allocation of the indirect costs (overhead) to the cost centers
3. calculation of the overhead charge
4. distribution of the overhead costs using the overhead charge percentages.

The idea behind cost allocation and the choice of bases for the overhead charge is that costs are allocated to each product based on the quantity of resources the product consumes. Direct costs are often used as bases for the overhead charge, as in the Trebla example. Of course, such allocations should not be made automatically; when better allocations bases are identified, naturally they should be used. Manufacturing overhead costs, for example, are usually allocated on the basis of direct labor costs (as a percentage of direct labor) but can also be allocated on the basis of direct work time (in SEK per labor hour) or on the basis of machine time (in SEK per machine hour), or using a combination of these methods. Materials overhead costs are usually allocated on the basis of direct materials (a percentage of direct materials). In certain situations, however, it may be better to allocate materials overhead costs, or some portion of them, on the basis of how much material is used (in SEK per kilogram, SEK per square meter, etc.).

It is also important to avoid making these cost allocations to cost centers automatically. The allocations often require several steps: costs are allocated to the organization unit (cost center) where they occurred, and then further allocated using the allocation method (described above) to the cost units. Costs should, of course, be allocated to the cost centers where the bases for the overhead charge best reflect how the resources are consumed.

In order that their calculations and allocations are as accurate as possible, the people who do this work should have an in-depth understanding of the relevant activities.

When a company's overhead costs are allocated routinely to cost units, as in full costing using the absorption method, situations will always arise where costs are misallocated. For this reason, full costing is often criticized. We will come back to this issue in Section 4.6.

4.5 Step costing

Researchers and practitioners have debated the advantages and disadvantages of full costing (e.g., using the absorption method) and contribution costing for years. In practice, it appears that companies often use both costing methods (although on different occasions) or in combination. *Step costing* tries to combine the advantages of the complete cost breakdown of full costing and the contribution margin analysis of contribution costing by gradually calculating the surplus in steps. (Not to be confused with step costs, a type of fixed costs, which was discussed above.) In our step costing example, the four steps are identified as CM1 (contribution margin 1), CM2, CM3, and CM4. The first step, CM1, corresponds to the contribution margin in the contribution method, whereas the last step, CM4, corresponds to the difference between total revenues and total costs (i.e., the same profit that the full costing method produces).

EXAMPLE

The Beverage Division is part of the global company SADIDA that mainly manufactures and sells perfume, clothes, sunglasses, and watches. The company's Beverage Division makes and sells two sports drinks (Blue and Green) and two fizzy drinks (Black and Yellow). Step costing (in four steps) for the beverages (per liter) is as follows:

	Sports drinks			Fizzy drinks			Beverage total
	Blue	Green	Total	Black	Yellow	Total	
Specific revenue (/liter)	SEK 15	SEK 19		SEK 50	SEK 70		
Specific costs (/liter)	SEK −7	SEK −8		SEK −25	SEK −30		
CM1 (/liter)	SEK 8	SEK 11		SEK 25	SEK 40		
Quantity	750,000 liter	330,000 liter		220,000 liter	40,000 liter		
–							
CM2	MSEK 6	MSEK 3.63	MSEK 9.63	MSEK 5.5	MSEK 1.6	MSEK 7.1	
Product group's specific costs			MSEK −2.43			MSEK −0.59	
–							
CM3			MSEK 7.2			MSEK 6.51	MSEK 13.71
Division's specific costs							MSEK −8.5
CM4							MSEK 5.21

CM1 is the *contribution margin per product* (per liter) calculated by contribution costing. CM2 is the *total contribution margin per product* – that is, CM1 multiplied by volume. CM3 is the total contribution margin for the product groups (in this example, CM2 from the sports drinks less their product group's specific costs plus CM2 from the fizzy drinks less their product group's specific costs). Finally, CM4 is the total contribution margin that the Beverage Division makes to the company (i.e., both product groups' CM3 less the Division's specific costs). Of course, the structuring of the contribution margins into different levels can be handled differently if it better suits the corporation.

The treatment of the specific cost concept in step costing can be confusing. The specific costs at a particular level of the company are often common costs at a lower level. As an example, let us look at the calculation for the two sports drinks (Blue and Green). The corporation has revenue of SEK 15 per liter sold of Blue and revenue of SEK 19 per liter sold of Green. Specific costs are SEK 7 per liter of Blue and SEK 8 per liter of Green. CM1 for the two drinks is as follows: Blue, SEK 8 per liter, and Green, SEK 11 per liter.

Continuing with the example, CM2 is then the total contribution margin from the two sports drinks (i.e., CM1 multiplied by volume for each drink). To obtain CM3, we reduce CM2 by the product group's specific costs. These costs are the common costs if we look at them from a product level, which means that the company would have these costs even if it stopped manufacturing the sports drinks. However, from the company's point of view, these costs are specific costs, belonging to the sports drinks. Thus, if the company ceased manufacturing sports drinks, it would incur no costs for this product group.

Adding the CM3 of both groups – sports drinks and fizzy drinks – and subtracting the Division's specific costs, gives CM4. The Beverage Division's specific costs are common costs for the two product groups of sports drinks and fizzy drinks. This means that the Beverage Division would have these costs even if a product group were discontinued. For the company, however, these costs are specific costs; the company can decide to close the whole Beverage Division.

Using step costing, as illustrated here, a company can focus on each product, the product groups, and division profitability. In classic contribution costing, only product profitability is calculated.

4.6 Activity-based costing (ABC)

Activity-based costing (ABC) has emerged as a reaction to the criticism of the traditional costing allocations. The main criticism is that those allocation methods treat too many costs as indirect costs. As discussed above, direct costs are traced directly to the *cost units*, while indirect costs are assigned to *cost centers* (often departments) and then allocated to the cost units (e.g., products). This allocation is made using percentage rates on predetermined bases for overhead charges. In most industrial companies, the bases for overhead charges are almost always related to production volume. This means that all indirect costs, in the allocation calculations, are treated as if they vary with production volume. Of course, there are costs that are not directly affected by volume changes. Examples are costs related to plant facilities, machines, and managerial activities. These costs are unaffected by small volume changes.

In ABC, cost concepts such as fixed/variable or direct/indirect are not in focus. Instead, ABC is based on a company's *activities* rather than on its departments. The idea is that costs are incurred when these activities are performed, not by the organizational units. Cost allocations are made after mapping these company activities and determining their costs. See Figure 4.6 for an illustration of ABC.

In ABC, you look at what is concealed behind the indirect costs. The activities are mapped, as well as the reasons for their amounts, at each cost center. For example, the mapping can reveal that the costs of the operations planning are determined by the number of manufacturing orders, that the costs of making engineering changes are determined by the number of change orders, and so on. These dimensioning factors are called *cost drivers* in ABC.

Figure 4.6 Activity-based costing allocation principles.

EXAMPLE

Let's return to Trebla AB. Several years have passed. Both Easy Chair and Office Chair are still in production along with many other products. Among the new products are Junior and Senior (also chairs) that are part of a new marketing initiative. The following production facts are available for Junior and Senior:

In the manufacturing process, it takes 0.05 hours of direct labor per chair (the same for both chairs). Factory wages are SEK 200 per hour. Thus, each chair costs SEK 10 in direct labor.

Annual production volume: 1,000 Junior chairs and 25,000 Senior chairs.

Total direct labor per year for the whole company is 10,000 hours. Total indirect manufacturing costs (manufacturing overhead) for the company are SEK 10 million per year.

How much of these indirect costs should each chair carry?

$$\text{MgO charge} = \frac{\text{SEK 10 million}}{10,000 \text{ hours}} = \text{SEK 1,000 per hour}$$

or

$$\text{MgO charge} = \frac{\text{SEK 10 million}}{(10,000 \text{ hours} \cdot \text{SEK 200 per hour})} = 500 \% \text{ of direct labor}$$

This gives us the following calculation of manufacturing costs per chair (all figures are shown in SEK):

	Junior	Senior
Direct labor	10	10
Indirect manufacturing costs	50	50 (500% of direct labor)
Total	60	60

These traditional full cost calculations show that Junior and Senior cost the company the same for direct labor and manufacturing overhead.

EXAMPLE

In our example, an investigation reveals that indirect manufacturing costs (overhead) are caused by the following cost drivers:

- 50 % of MgO is caused by the cost driver *number of direct labor hours*
- 30 % of MgO is caused by the cost driver *number of manufacturing orders*
- 20 % of MgO is caused by the cost driver *number of items (of the chairs)*

[yr = year; no = number of items]
We also assume the following information on the cost drivers:

	Junior	Senior	Total
Direct labor hours	0.05 hr/unit	0.05 hr/unit	10,000 hr/yr
Manufacturing orders	20 orders/yr	125 orders/yr	1,000 orders/yr
Number of items	1 unit	1 unit	80 units
Year's total	1,000 units.	25,000 units	

With this information, we can now calculate the costs per unit for the three cost drivers:

Cost driver		Cost/unit
Labor hours	(50 % of SEK 10 M)/10,000	SEK 500 per hr
Manufacturing orders	(30 % of SEK 10 M)/1,000	SEK 3,000 per order
Item numbers	(20 % of SEK 10 M)/80	SEK 25,000 per no and yr

Calculations for Junior and Senior are as follows:

Junior Activity		Cost/unit
Direct labor	0.05 hr · SEK 200	SEK 10
Labor hours	0.05 unit · SEK 500	SEK 25
Manufacturing orders	20 units · SEK 3,000 per 1,000 units	SEK 60
Item numbers	1 unit · SEK 25,000 per 1,000 units	SEK 25
Total		SEK 120

Senior Activity		Cost/unit
Direct labor	0.05 hr · SEK 200	SEK 10
Labor hours	0.05 unit · SEK 500	SEK 25
Manufacturing orders	125 units · SEK 3,000 per 25,000 units	SEK 15
Item numbers	1 unit · SEK 25,000 per 25,000 units	SEK 1
Total		SEK 51

The traditional full cost calculations revealed that the products cost the same. However, by applying ABC, the investigation shows the activities that incur costs and how the different products relate to these costs. Junior is 135 % more expensive than Senior. If the sales price were based on the traditional calculation method, Senior would have subsidized Junior. The ABC calculation reveals the significant difference in the cost of the two chairs.

The ABC example shows that when manufacturing overhead is allocated to products based on activity volumes, certain products may often subsidize others. High-volume products carry a disproportionate share of the indirect costs compared to low-volume products. Standard products subsidize non-standard products, products that are manufactured by old machines subsidize new products manufactured by special machines, large-volume customers subsidize small-volume customers, and so on. The explanation is that in traditional overhead calculations, manufacturing overhead costs are allocated to all products by the same percentages.

In practice, however, it is difficult to make a complete breakdown of the costs of all activities and their cost drivers. Companies that have introduced ABC have often retained parts of their direct costs allocations. See Figure 4.7. In terms of indirect costs (overhead), companies are often content to determine costs of introducing new products and costs of handling orders by allowing, for example, the cost drivers of item numbers and order lines to carry these costs. The remaining costs are often allocated in the traditional way.

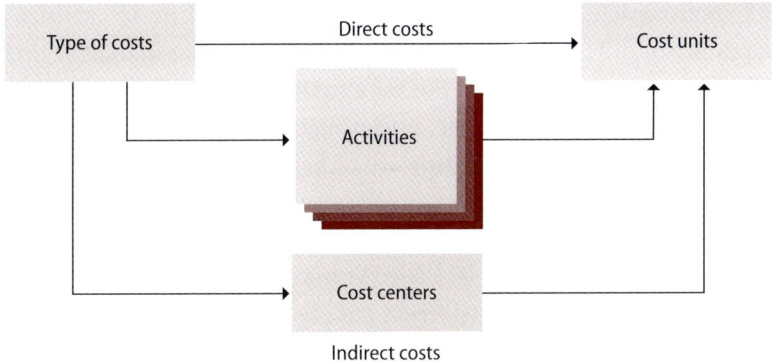

Figure 4.7 Activity-based costing in practice.

There is another criticism of ABC. Critics claim there are no significant differences between traditional costing methods and ABC. It is claimed that the only real difference is that ABC has introduced more cost allocation bases. In this view, ABC is merely a refined method in which more care is taken in the allocation of the indirect costs.

4.7 Operations determine the costing techniques

The chapter's introduction explained that product costing techniques emerged in response to companies' wishes to calculate the cost to manufacture products and thereby to set product prices. However, the complexity of these calculations increases when companies begin to produce many different products. Historically, companies made such calculations for the entire company or factory rather than for individual products.

Today, there are many situations in which it is necessary to make cost calculations: to prepare tenders to customers; to analyze efficiency of different operations processes, work methods, or manufacturing equipment; to allocate costs to various products; or to prepare calculations for investment decisions. (See Chapter 5 for a discussion of investment calculations). Thus, calculations can be made using various types of objects in focus.

It is the decision situation that determines which cost unit is the most logical for the cost in focus. This means that in every situation you should think about which cost unit to use. Companies may want to calculate the profitability of different geographic markets, of different market segments, or of different product designs. For many activities, time is the appropriate cost allocation base. For example, a printing company may choose machine hours because it sells printing capacity rather than products. A consulting firm may choose hours because it invoices customers by the hour.

It can sometimes be difficult to determine the most appropriate cost allocation base. When, for example, a telecommunications company sets prices, many questions arise. Are customers paying for the ability to make calls? If so, the company should charge all customers a fixed fee. Are customers paying for the length of their calls? If so, the company should set prices based on the time customers spend using the telephone. Or should the company set prices based on the broadband speed or transmission capacity it provides? And, of course, which competing alternatives are available to customers? These are difficult questions that should be considered in deciding on a cost allocation base.

Which cost units are appropriate for different types of operations? And when should they be used? These questions are another way of formulating and thinking about the various issues discussed above. Of course, the assumption is that the costing methods chosen reflect a company's operations and different activities. Yet another consideration may be relevant – costing methods influence activities and employee behavior at the company. For example, in some situations, internal resources (e.g., machines) are so expensive that units within the company may decide to purchase products produced by these resources from external companies rather than producing them in-house. Then, when fewer products use the internal resources, costs are allocated to fewer products. These products then have a higher cost, reducing their profitability. Next, those responsible for these products also consider looking outside the company for cheaper alternatives. The situation becomes a downward spiral as people in charge of the machines complain their machines are not used merely because of the allocation principles used.

An example of this deteriorating situation was when some universities in Sweden changed the principles behind handling the costs of facilities – from allocating these costs to a central university function to allocating the costs to the various departments and institutions based on how much floor space they used. Overnight, research activities that required large floor space became more expensive than other research activities that required less floor space. This outcome affected departments with large laboratory spaces. Thus, because research in these areas was now more expensive, fewer researchers used these facilities; instead, many researchers turned to research projects driven primarily by theoretical simulations and which required little use of the physical facilities. Consequently, the research priorities were controlled by rental costs rather than by scientific needs.

A fundamental idea behind product costing is the need to gather information on decision-making alternatives. This means that the way in which allocations are made will influence a company's operations. For example, what can be measured will be performed. Thus, costing methods should be designed so that their intended purposes are taken into consideration. Among other things, product costing is closely related to the nature of a company's operations and to its objectives.

4.8 Calculations – before and after

One of the main uses of product costing is the calculation of costs in advance of a decision or an activity: *cost estimates*. To check for the accuracy of such calculations, however, many companies also conduct *follow-up costing* calculations after the activity is completed.

Follow-up costing requires information that is available in a company's cost accounting system (e.g., the expenditures for purchased raw materials, for salaries, and production facilities). See Chapters 6 and 7 for a discussion of accounting systems. Thus, the cost accounting system provides data on the resource consumption, or, in actuality, the accounted value of the resources consumed during the period. For an accurate allocation of these costs, more detailed cost accounting is required than is mandated by law. In addition, efficiency analyses also require data on company performance for the period under examination.

EXAMPLE

When Trebla checked on how much raw material was used in a period, the cost accounting records provided the following data:

- The manufacture of Easy Chairs used materials that cost SEK 1,275,000.
- The manufacture of Office Chairs used materials that cost SEK 2,090,000.

However, these figures tell us nothing about the efficiency of the resource utilization. For this calculation, we need to know how many Easy Chairs and how many Office Chairs were produced – that is, the amount of materials that should have been used to produce this number of chairs. According to our estimates, each Easy Chair required materials at a cost of SEK 40, and each Office Chair required materials at a cost of SEK 60. Records show that in this period the company manufactured 30,000 Easy Chairs and 35,000 Office Chairs.

Now we can make a follow-up cost comparison between the estimated cost and the actual cost of the chairs. All figures are shown in SEK.

	Easy	Total	Office	Total
Estimate of materials usage	30,000 units · 40 per unit	1,200,000	35,000 units · 60 per unit	2,100,000
Actual materials usage		1,275,000		2,090,000
Difference		−75,000		10,000

The follow-up calculation reveals that the manufacture of Easy Chairs used materials costing SEK 75,000 more than estimated while the manufacture of Office Chairs used materials costing SEK 10,000 less than estimated. There are several possible explanations for these differences. Possibly there were production problems – untrained staff, machinery malfunction, etc. – with Easy Chairs, resulting in materials damage or waste. However, another explanation may be that the estimated cost was incorrect because of overly optimistic estimates on materials usage or the time required to complete the manufacturing. The follow-up costing helps focus the company on improving its cost estimates.

CHAPTER 5

Capital investment analysis

A vital issue for companies, for their long-term viability, is the need for resources that allow them to compete in the market. A newly founded company needs initial resources (e.g., capital). A manufacturing company that is planning to increase its production capacity needs to determine how many new machines are required. Another company that is evaluating whether to enter a new market must think about the investments needed for new production facilities and for compliance with new rules and regulations (e.g., emission standards). Or another company may be weighing the costs required for new staff training programs or for a new manufacturing platform (e.g., a new mine or a new paper machine). This chapter addresses the principal conditions surrounding companies' considerations for such long-term *investments*.

5.1 What is a capital investment?

Many different activities can be described as investments. Theoretically, an investment means the opposite of consumption (i.e., resources are created, not consumed). For an industrial company, investing usually means a significant monetary transaction, such as the acquisition of a physical resource like machinery, production facilities, or a major IT system. However, investing may also mean initiating a new marketing campaign, starting a research and development project, or purchasing staff training programs. In the financial markets, investing means, for example, buying and selling of shares, options, or other securities. In practice, there is no single formal definition of an investment; however, an investment is usually related to major financial commitments over a long time period. Frequently, the company and/or industry traditions and circumstances determine if a major expenditure is an investment or not. For the independent taxi driver, a new car purchase is a significant investment; a large company may regard its new car purchase as just another, routine business expenditure – one among many.

The commonality among these examples is that they all require the present outlay of money intended to produce a positive result in the future. Thus, an investment is by definition a major financial undertaking that will have long-term consequences. Because investments involve resource consumption, the established techniques for investment calculation are strictly built on the specific inflows and outflows of money – the *receipts* and *payments* – related to a particular investment.

Investments

An investment is a capital outlay that has payment consequences over an extended time period, at least more than a year.

Receipts and Payments

A **receipt** (cash in) occurs at the moment when money is received (e.g., when a customer pays an invoice for goods or services the company has delivered).

A **payment** (cash out) occurs at the moment when money is paid (e.g., when the company pays an invoice for the goods or services the supplier has delivered).

Making investment appraisals (investment calculations) means looking at the company's *cash flows* (in and out). In making an investment, the investor pays out money now in the expectation that money will be received later (in the amount of the original investment plus more). Because investments, as defined above, are long term, the investor needs to know (estimate) how much money will eventually be received and when. By including all relevant receipts (cash in) and payments (cash out) in the investment appraisal, it is possible to compare different investment alternatives.

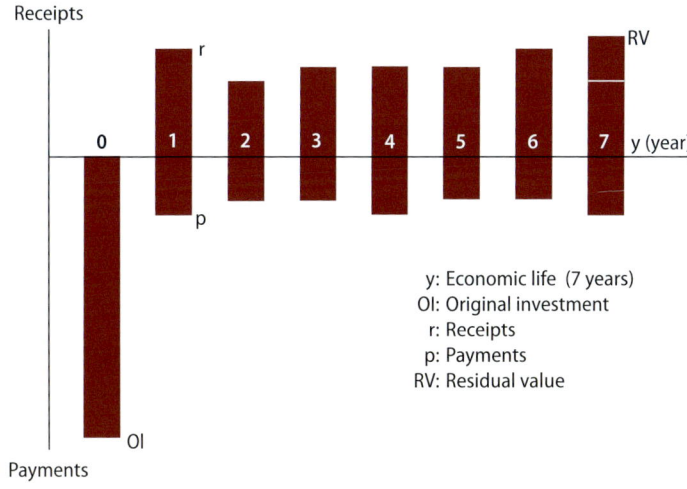

Figure 5.1 Investments illustrated as cash-flow charts.

In the appraisal, we include only those receipts and payments following the original investment (payment). These receipts and payments are often illustrated using the *cash flow chart*. See Figure 5.1. Typically, a large payment (cash outflow) occurs initially, for example for a new machine. The cash inflows are the receipts as the result of the investment: increased sales, for example. At the end of the machine's life, when it is sold for its residual value in the last year, another receipt (cash inflow) results.

The cash-flow chart shows the investment payments on a timeline. Receipts (r) are shown above the horizontal line and Payments (p) are shown below the horizontal line. The *Original Investment* (OI) is the initial cash payment, marked by the downward bar at the end of Year 0 (i.e., at the beginning of Year 1). The future receipts (cash inflows) and payments (cash outflows) are graphed by period (usually yearly) with upward or downward bars. A running calculation of the receipts and payments can be summarized by a single bar that shows the difference between them for each period. We can calculate the *over* (or *under*) *cash surplus* for each year.

The *Residual Value* (RV), that is, the amount the investment is worth at the end of its economic life (y), is also shown as a bar. A positive residual value is estimated, for example, when the company expects to take the machine out of service and sell it. A negative residual value is estimated when it is assumed the cost of disposing of the machine will be higher than its estimated selling price. Possibly the company can sell the machine in a second-hand market; possibly the company will have to pay to have the machine removed.

It is important to differentiate between *economic life* and *technical life*. Economic life, the period in the cash-flow chart, refers to the time period until other investment alternatives appear more profitable. In other words, continuing with the machine example, it is assumed that at the end of the machine's economic life, it will be more profitable to make a new investment by replacing the original machine. Technical life refers to the machine's total functional time, limited only by its physical and technical capacities. A machine that is still functional, for example, may be so expensive to maintain that it is more profitable to buy a new one. In some instances, technology develops so rapidly that a company may decide that even a relatively new machine is no longer profitable compared with new models that require less energy or fewer repairs. Because of maintenance issues and technological developments, economic lives are often considerable shorter than technical lives. By definition, economic life cannot be longer than the technical life. Computers are a good example. Many companies replace them long before they cease working. In many practical instances, however, because it is quite difficult to imagine future technological developments, it is also difficult to estimate an investment's economic life.

Economic life

The total time in which the investment achieves its maximum profitability. This is the time in which investment calculations are made.

Technical life

The total functional time limited only by the investment's physical and technical capacities. Technical life exceeds or is the same as economic life.

Frequently, however, machines are used in production long after the end of their economic lives or after they have been fully depreciated in the accounting records. See Chapter 7. Depreciation is a bookkeeping calculation only, and depreciable life does not necessarily coincide with economic life.

In practice, investment calculations often use standard depreciation periods as a reasonable approximation of economic lives. This calculation, however, should not be confused with actual *depreciation*, which is a bookkeeping/accounting event.

5.2 How are investments evaluated?

In the discussion of investments, it is important to remember that any calculation is only a part of the comprehensive *investment evaluation* required. The reason is that many different factors, even those that cannot be calculated in monetary terms, must be considered.

A fundamental difficulty in making investment calculations is that many factors are often unquantifiable. How can factory improvements designed to improve air or water quality be evaluated monetarily? What is the monetary outcome of investing in a new IT system? How do we quantify the results of initiating new marketing programs or of opening new sales channels? Take the example of an investment in a new mine: besides the cost of the original investment, how much can the company expect to pay for related costs, such as the cost of the regulatory evaluations of the environmental impact?

In many cases, a company may make a large investment for strategic reasons even though it is impossible to estimate the short-term economic profit. It is, of course, very difficult to quantify the long-term effect on profitability of some investments. In such cases, investment evaluations require fairly extensive research of, for example, market trends, technology developments, and competitors' actions. The final investment decision, then, requires many types of evidence besides the financial data that the formal calculations provide (which are presented in later sections of this chapter).

In practice, an investment decision, whether for a small or large investment, almost always involves some factors and considerations that are not easily quantified. It is the unfortunate situation that quantitative calculations are sometimes over-emphasized. What can be measured in numbers is usually prioritized.

5.3 Investment calculations

Investment calculations are an important part, but not the only part, of investment evaluations. The calculations deal with how investments influence the cash flows (receipts and payments) – in other words, how much money moves in and out of the company. These calculations answer questions such as the following. How much will we reduce manufacturing costs if we invest in the latest machine model? How much will sales increase if we invest SEK 20 million in a distribution network in Poland? How much is the financial return if we procure new mining equipment or a new paper machine? As stated above, although such monetary calculations are an essential part of investment evaluations, they are not the only considerations in making investments.

The primary purpose of investment calculations is to provide a basis by which to evaluate the profitability of an investment project. For companies with several investment alternatives, these calculations also allow comparisons and ranking of the alternatives. On the basis of their expected profitability, the company can decide which investments should be made. As a simple example, in Figure 5.2, we present two alternative investments: A and B. The question: Which of the two investments is more profitable?

It is very simple to see that Alternative A is a better investment than Alternative B. The original investment (the cash outflow or payment at the

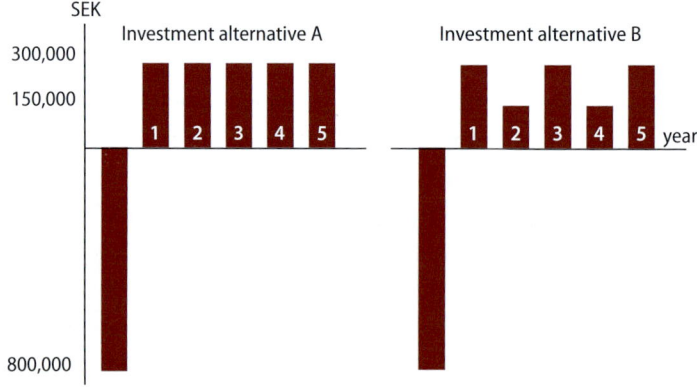

Figure 5.2 Two capital investment alternatives: A and B.

beginning of Year 1) is the same (SEK 800,000). Alternative A's receipts (cash inflows) in the five years are either larger or the same as Alternative B's receipts. By calculating the results in monetary terms over the economic lives (five years in both alternatives), we are able to compare and rank the two alternatives. The cash-flow charts tell us that A is a better investment than B. However one question remains: Are the investments profitable? Should the investor choose A, neither A nor B, or perhaps both A and B?

To answer this question, we must know the investor's *required rate of return* on investments. This rate of return can be expressed in various ways depending on which investment calculation method is used. Often the *cost of capital* is used to specify the required rate of return. Another possibility is to specify the longest acceptable time period, the *payback time*, needed to repay the original investment. If the required payback time, in our example, is three years, A is profitable, but B is not. It takes slightly less than three years for A to repay the original investment but more than three years for B. However, if the required payback time is four years, both alternatives are profitable investments. If the required payback time is two years, neither A nor B is a profitable investment. Later in the chapter we will discuss how the cost of capital and the payback time are determined. For now, we present the four most commonly used methods to make investment calculations: the net present value method, the annuity method, the internal rate of return method, and the payback method.

Net present value method

The *net present value method* is an interest rate calculation method. Using a predetermined interest rate (the cost of capital), the investment alternatives' receipts and payments are discounted to the same time point – the beginning of the original investment. One says the future payments are "discounted to present time" using the *cost of capital* as the discount (interest) rate. At the same time that the cost of capital expresses the required rate of return, it also expresses the *time preference*. This means that a future receipt/payment is considered less valuable than the same receipt/payment today. If we have money today, we can, as an alternative to the capital investment, save it and earn interest from a bank account or invest it in the current operations and earn short-term returns. Thus, the use of the interest rate called the *cost of capital* is a way to take the long-term nature of investment decisions into account. Furthermore, a capital investment sometimes requires loans on the financial market, and this means that the investment's rate of return must exceed the costs related to financing (capital costs) in order to be profitable.

Time preference

A receipt we receive today is worth more than the same receipt at some future date since today's receipt can be used in the current operations and can earn short-term returns.

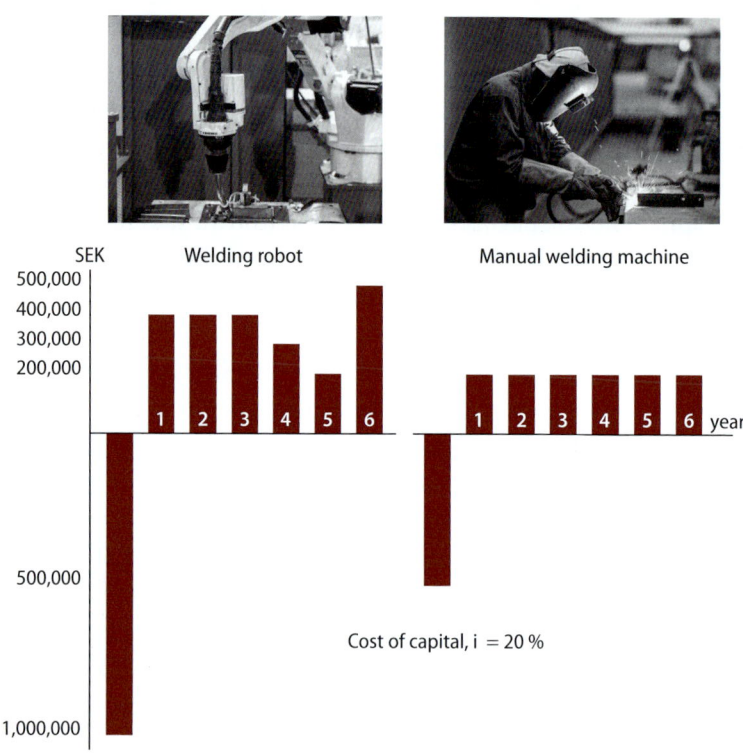

Figure 5.3 Two investment alternatives: a welding robot and a manual welding machine. Photo sources: Amnarj Tanongrattana/shutterstock.com (robot) and Sasin T/shutterstock.com (manual machine).

Let us take an example of the net present value method. The company "Industrial Components" specializes in the manufacture of welded products and components. Turnover has increased significantly in recent years. Management, which is investigating different expansion possibilities, asks the factory manager to report on the various manufacturing investments needed. The factory manager reports that a new welding machine is needed. After looking at alternatives, the factory manager must choose between a welding robot and a manual welding machine. The welding robot is more expensive, but does not require a manual operator. The manual welding machine is less expensive, but requires a manual operator. Figure 5.3 shows two cash-flow charts that present the cash flows (cash in/receipts and cash out/payments) for the two alternatives. Management has set the cost of capital (the required rate of return) at 20 %. Which alternative is better?

Using the cost of capital as the interest (discount) rate, we can discount the receipts back to the present time (also referred to as time zero), that is, to the beginning of Year 1. The principles for discounting receipts and payments are simple and follow the usual interest calculation rules. The simplest way to demonstrate this calculation is to calculate amounts forward in time. Suppose we have the opportunity to receive a sum of money today (X) or some amount of money in one year. The question is the following: how much money must we receive one year from now that is equal to the money we receive today? Suppose also that the amount received today can be invested so that it earns the same rate of interest as the cost of capital, say i percent. After one year, we will have our original amount of money plus the accumulated interest: $X \cdot (1 + i)$. Receiving that future sum is the same as receiving X today.

Cost of capital

The cost of capital expresses the required rate of return on an investment and is used in investment calculations for receipts and payments over time.

The cost of capital is decided internally in the company.

For different kinds of investments, different costs of capital are used.

If X is SEK 1,000 and the cost of capital is 20 %, the corresponding amount in one year is SEK 1,200. This amount is called the *future value*. The amount can be calculated forward any number of years (y) by multiplying it by the *future value factor*. The future value mathematical formula, which is the normal interest-on-interest formula, is shown in the box.

Future value factor formula

$$FV = (1 + i)^y$$

In a similar calculation, receipts and payments can be discounted backward in time. An amount we receive in a year divided by the FV formula can be calculated to the present time, which is referred to as its *present value*. The mathematical calculation for discounting is thus the inverse of the FV formula. Discounting an amount backward in time for a number of years is quite simple using the present value formula.

Present value factor formula

$$PV = \frac{1}{(1 + i)^y}$$

Now we can make an investment calculation using the present value method. The calculation is simplified, however, when we introduce the *cumulative present value*. With this formula, in a single step we can calculate the present value of a series of equal receipts (or payments), spaced at equal distances.

Cumulative present value factor formula

$$CPV = \frac{1 - (1 + i)^{-y}}{i}$$

Formula for Net Present Value in cases when there is an equal payment surplus each year

Net Present Value = OI + a · CPV (i %, y year) + RV · PV (i %, y year)

OI = Original investment (usually negative)
a = annual surplus, i.e. r – p (same each year)
CPV = cumulative present value factor
i= cost of capital (discount rate)
y = economic life
RV = residual value
PV = present value factor

These factors are presented in table form for different interest rates and different time periods. You will find these tables at the end of this book. There are also software programs available for these calculations.

EXAMPLE

Net Present Value calculation:
Welding robot or manual welding machine
We begin with the welding robot investment. First we subtract the Original Investment (OI). Then, for the first three years, we use the cumulative present value calculation (CPV) to discount the three equal receipts of SEK 400 each year. In Years 4 to 6, in which the amounts are unequal, we discount the receipts using present value calculations (PV).

Net Present Value = OI + a_{1-3} · CPV(20 %, 3 yr) + a_4 · PV(20 %, 4 yr) + a_5 · PV(20 %, 5 yr) + a_6 · PV(20 %, 6 yr)

CPV (20 %, 3 yr) = 2.1065
PV (20 %, 4 yr) = 0.4823
PV (20 %, 5 yr) = 0.4019
PV (20 %, 6 yr) = 0.3349

The net present value for the welding robot alternative is:

$-1,000 + 400 · 2.1065 + 300 · 0.4823 + 200 · 0.4019 + 500 · 0.3349 \approx$ SEK 235,000

The calculation for the manual welding machine is simpler. First we subtract the Original Investment (OI). Then we can use the CPV factor to discount all the equal payments to the present time in one step.

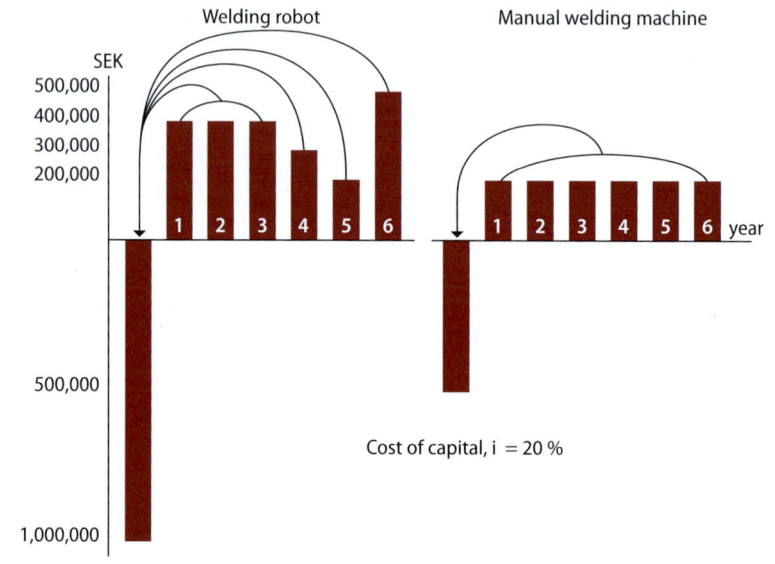

Net Present Value = OI + a · CPV(20 %, 6 yr)

CPV (20 %, 6 yr)= 3.3255

The Net Present Value for the manual welding machine alternative is:

−500 + 200 · 3.3255 ≈ SEK 165,000

While the results of the calculations for both machines are positive, the net present value is larger for the welding robot. Both investments return over 20 % (because the net present value is greater than zero), but the welding robot is a better investment than the manual welding machine.

Decision rules for the Net Present Value Method

An investment with a present value greater than the original investment is a profitable investment. The difference is called the *net present value* and should be greater than zero.

The investment alternative with the largest net present value is the most profitable.

It is often the case that a company's access to capital limits its ability to make investments. In practice, different investments compete for capital. Therefore, it is useful to examine how well different investments use the invested capital. This calculation is possible using the *net present ratio*. The ratio is simply the net present value in relation to the original investment.

Net Present Ratio

$$\text{Net Present Ratio} = \frac{\text{Net Present Value}}{\text{Original investment}}$$

A high net present ratio is better than a low one. In the example we have used to this point, the welding robot has a net present ratio of 0.24. The manual welding machine has a net present ratio of 0.33.

EXAMPLE

Net Present Ratio: Welding robot or manual welding machine

Net Present Ratio for the welding robot is: $\dfrac{235}{1\,000} \approx 0.24$

Net Present Ratio for the manual welding machine is: $\dfrac{165}{500} = 0.33$

Thus, the manual welding machine makes better use of capital invested even though the net present value calculation shows that it is the worse alternative. Another way to state this is that the manual welding machine can produce a higher rate of return than the welding robot, which is also the result if the two alternatives are calculated using the *internal rate of return* method (see below).

Although the net present value method is not as simple as the payback method (see below), it is not particularly difficult. Moreover, it directly takes the time preference into account with its use of the cost of capital interest rate. However, a disadvantage of the net present value method is that the method should not be used to compare investments with different economic lives.

If we know for certain the amount of the receipts (cash in) and the amount of the payments (cash out) in perpetuity (forever) for the investments, the net present value method can be used without special consideration even if the economic lives of the alternatives differ. In practice, however, this is difficult to know because you must know what will occur after the end of the investment's economic life. Will we reinvest in or significantly change the activities related to the investment? If we reinvest, how much should we reinvest? If we know that the investment will be discontinued, there is no concern because all results appear in our calculation.

One way to deal with this problem and still use the net present value method is to shorten the life of the alternative that has the longer economic life. Thus, you simply calculate two alternative investments over the same time period: the shorter economic life. To compensate for the years removed from the calculation for the investment with the longer economic life, you use a fictitious residual value for the final year. Then the net present value calculation is performed in the normal way. The residual value becomes an estimate of the receipts not received (or payments not paid) when the longer economic life is shortened.

Another consideration is that the size of the discount rate (i.e., cost of capital) can influence the ranking of the alternative investments. In a simplified way, you can say an alternative with a larger original investment (like the welding robot in the example) benefits if the discount rate is low. Conversely, the alternative investment with a smaller original investment benefits if the discount rate is high. The same result is possible if the peaks of the future payment surpluses occur at different points in the future for the two alternatives. It is expected then that the size of the discount rate influences the ranking because the discount rate in fact measures how much the future payments are discounted.

Annuity method

In the *annuity method* the original investment and all future receipts (cash in) and payments (cash out) are distributed over the investment's economic life in equal annual amounts. Such future receipts and payments are called *annuities*. See Figure 5.4 for an example. A simple comparison is to use annuity loans that banks and other financial institutions make. With such loans, the borrower pays the lender an equal monthly amount consisting of interest payments and loan principal payments. Thus, during the entire loan period, the borrower pays the same sum each month. The average monthly payment is calculated based on the total sum. Stating this calculation as an average is somewhat erroneous because the calculation uses the cost of capital as the discount rate. In principle, the net present value method and the annuity method are two variations of the same calculation.

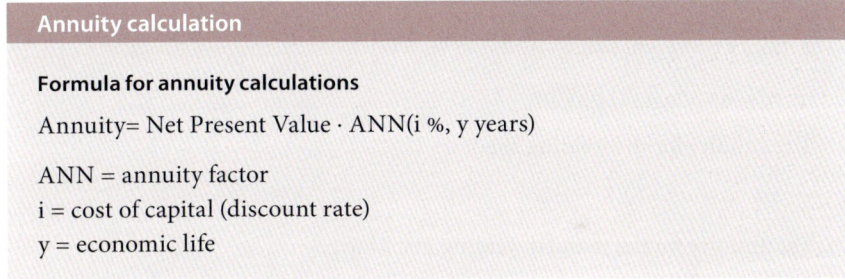

Annuity calculation

Formula for annuity calculations

Annuity= Net Present Value · ANN(i %, y years)

ANN = annuity factor
i = cost of capital (discount rate)
y = economic life

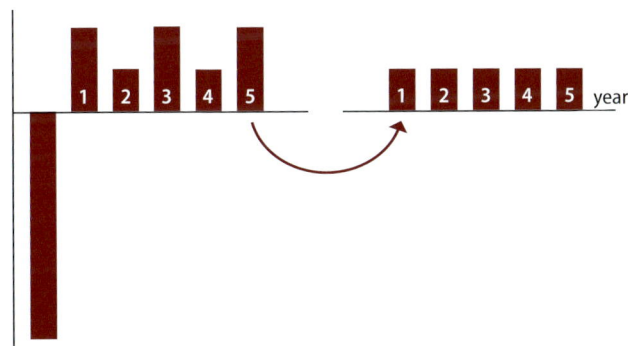

Figure 5.4 The annuity method calculates an "average annual" payment.

Typically, the calculation of an investment's annuity is made first by calculating the present value of the investment over its entire economic life. Then the present value is multiplied by the *annuity factor* (ANN), which is simply the inverse of the CPV. The annuity factor varies with the discount rate (i) and the economic life (y). The tables at the end of the book present these factors.

EXAMPLE

Annuity calculation: Welding robot or manual welding machine
Because we have already calculated the present value of both alternatives, the annuities can be easily calculated by multiplying the present value by the annuity factor (ANN). (An alternative calculation for the manual welding machine is to multiply its original investment by the annuity factor and then subtract this amount from SEK 200,000. This is possible because all the receipts in Years 1 to 6 are the same. In short, the payments are already annuities). The annuity factors are in the table at the end of the book.

Annuity = Present value · ANN(20 %, 6 yr)

   ANN(20 %, 6 yr) = 0.3007

The annuity for the welding robot is:

   235 · 0.3007 ≈ SEK 71,000

The annuity for the manual welding machine is:

   165 · 0.3007 ≈ SEK 50,000

The welding robot has a larger annuity than the manual welding machine. See Figure 5.5. The decision rules for the annuity method are the same as for the present value method.

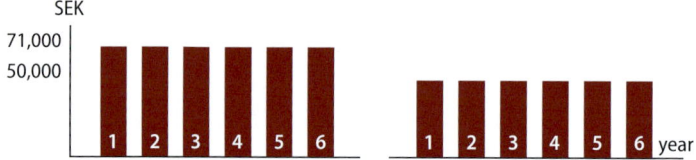

Figure 5.5 Annuity calculation: welding robot or manual welding machine.

Decision rules for the Annuity Method

An investment in which the annuity of the future receipts (and payments) is greater than zero is profitable.

The investment alternative with the largest annuity is the most profitable.

Thus, the welding robot appears the better investment although both investments are profitable because their annuities are greater than zero. If the investments, when compared, have the same economic life, the annuity method and the present value method will produce the same result.

With one exception, the annuity method, therefore, has the same advantages and disadvantages as the net present value method. The exception is that the annuity method can be used to compare alternatives with different economic lives if we assume *continuous replacement*. By continuous replacement, we mean that the operations in which the investment is made will continue after its economic life has ended. See Figure 5.6. For all the alternatives compared, a new investment is assumed at this point in time. If we

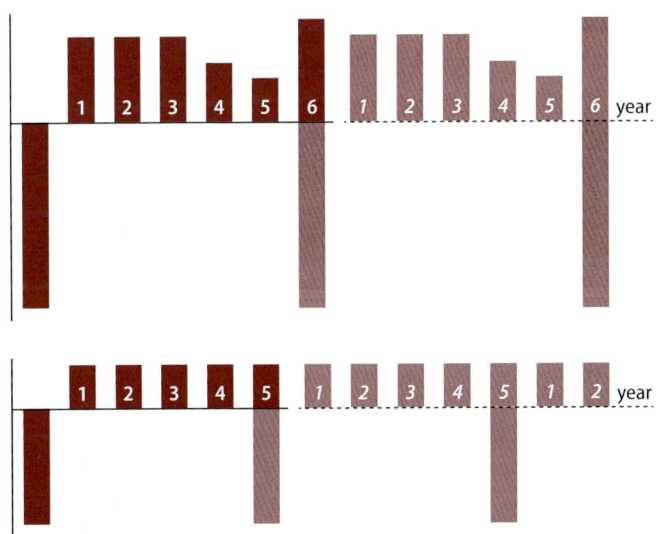

Figure 5.6 Continuous replacement.

assume that these future investments are the same as the investment made today, we can then make a chain of investments for the different alternatives. Because the annuity corresponds to an average year, we can calculate the first investment in each chain. But if we cannot assume continuous replacement, the annuity method can result in an incorrect decision.

In the same way that we use a ratio for the net present value method, we can use an *annuity ratio* to determine which investment provides the better return on the original investment.

Annuity Ratio

$$\text{Annuity Ratio} = \frac{\text{Annuity}}{\text{Original investment}}$$

EXAMPLE

Annuity Ratio: Welding robot or manual welding machine

Annuity Ratio for the welding robot is: $\dfrac{71}{1\ 000} = 0.071$

Annuity Ratio for the manual welding machine is: $\dfrac{50}{500} = 0.10$

Thus, the annuity ratio reveals that the manual welding machine provides the better return than the welding robot.

Internal rate of return method

The *internal rate of return* method calculates an interest rate of return (usually an annual rate) on an investment. This rate is called the *internal rate of return (IRR)*. In this method, you calculate the return on the investment in order to compare it to the required rate of return (i.e., the cost of capital). In the calculation, you calculate the interest rate at which the net present value of an investment equals zero. The net present value method and the annuity method begin with a predetermined interest rate (the cost of capital) in order to calculate the size of the surplus over the original investment discounted at that rate. In the internal rate of return method, the surplus is set to zero, and the interest rate that produces this result is the internal rate of return.

This rate can be calculated by trial and error, but that is rather laborious. Advanced calculators or spreadsheet software greatly simplify the calculation. Another, and more illustrative, calculation is depicted by a graph. The internal rates of return for the welding robot and the manual welding machine are graphed in Figure 5.7. Both curves show the present values plotted at different interest rates.

In Figure 5.7, we see that the welding robot has an internal rate of return of nearly 30 %, and the manual welding machine has an internal rate of return of around 33 %. The internal rate of return is the point where the curve crosses the X-axis. (Numerically, we can calculate the internal rates of return more exactly as 29.5 % and 32.7 %). Both interest rates are thus greater than the cost of capital (required rate of return) that was set at 20 %. This means that both alternatives are profitable. We had the same result with the net present value method and the annuity method.

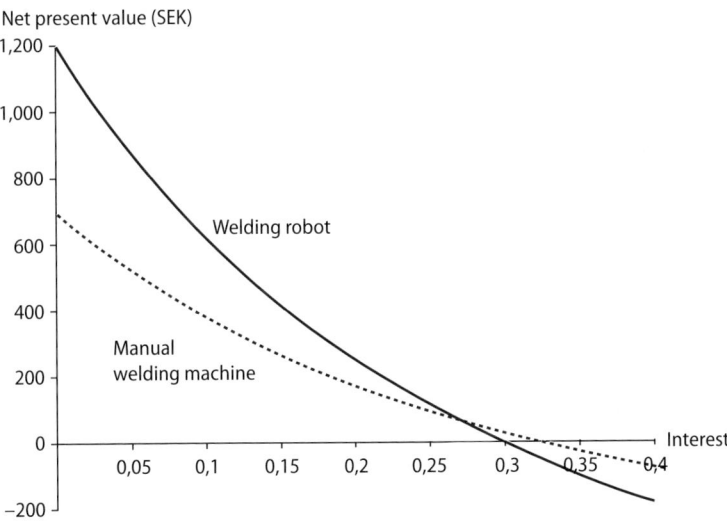

Figure 5.7 Graphic illustration of the Internal Rate of Return.

Decision rules for the Internal Rate of Return Method

An investment that has an internal rate of return that is greater than the cost of capital is profitable.

The investment alternative with the highest internal rate of return is the most profitable.

However, we see that the internal rate of return method case *does not* provide the same result as far as ranking the two alternatives. Using this method, the manual welding machine is more profitable than the welding robot, which is the opposite of the conclusions reached under the net present value method and the annuity method! This is consistent because the internal rate of return method is completely independent of the cost of capital. However, as Figure 5.7 shows, the net present values are equal at an interest rate of just over 25 % (the point where the curves intersect). At this rate, the net present value method and the annuity method provide the same conclusion. At higher interest rate calculations, the manual welding machine is the better choice, and at lower interest rate calculations, the welding robot is the better choice. Thus, the internal rate of return method gives a different result, when comparing the alternatives, than the other two methods. However, the results are the same as far as whether an investment is profitable or not.

Payback method

The *payback method* is the simplest of these investment calculations. According to the payback method, you simply determine the time it takes before the original investment is repaid (i.e., recovered). This period of time is called the *payback time*. As we have already seen, the decision criterion is expressed as a requirement for the repayment time (instead of as a required interest rate).

In our example, it takes two and a half years for both the welding robot and the manual welding machine to repay their original investments. Payback time is thus two and a half years for both alternatives; by this method, they are equally profitable.

The payback method does not tell us which alternative is preferable in terms of return on investment. However, its advantage is its simplicity. The calculation is intuitive and easy to understand. Therefore, it is a commonly used method. Another advantage is that indirectly the method takes the investor's uncertainty and *liquidity* into account. You only need be concerned with what happens to the investment before it is repaid. In addition, the payback method points to the investment with the shorter repayment period as the more profitable. A company can then use these funds to make other investments. An interest rate is not needed to discount the receipts (cash in).

However, the simplicity of the payback method is also a disadvantage. Time preference is expressed as the alternative that repays the investment the soonest, not by the value of the future receipts calculated using an interest rate. Another disadvantage is that the payback method does not deal with any considerations following the repayment period. If the alternatives have different consequences after the payments (cash out) are repaid, the payback method does not deal with these consequences.

Decision rules for the Payback Method

An investment that has a payback time that is shorter than the required repayment time is profitable.

The investment with the shortest payback time is the most profitable investment.

Summary of the methods

Table 5.1 Summary of the various investment calculation methods.

Method	Calculation	Decision rules	Advantages Disadvantages
Net present value (NPV)	All payments discounted back to time zero by using cost of capital interest rate	Profitable if NPV is positive The higher the NPV, the better	Explicit reference to time preference in use of cost of capital Doubtful with differences in economic lives (in comparisons)
Annuity (ANN)	All payments are distributed equally over the economic life by using cost of capital interest rate	Profitable if annuity is positive The higher the annuity, the better	Explicit reference to time preference in cost of capital Can be used with different lengths of economic lives, if continuous replacement is assumed
Internal rate of return (IRR)	Calculation of the interest rate that sets the investment's NPV to zero	Profitable if IRR is larger than the cost of capital The higher the IRR, the better (Can result in different ranking than NPV and ANN)	Enables direct comparisons of profitability between investments of different nature Complicated
Payback	Calculation of the time required to recover investment	Profitable if payback time is shorter than the required time The shorter the payback time, the better	Simplicity Tendency to over-emphasize short-term decisions

5.4 How is the required rate of return calculated?

Investment calculations are mathematically simple. None of the methods is particularly complex. Moreover, the calculations are greatly simplified by the use of the single sum and annuity tables, calculators, or computers. However, we should not be deceived by the simplicity of the calculations.

One of the greatest difficulties in investment calculations involves the *required rate of return* – that is, determining the cost of capital and the limit of the acceptable payback period. How do we set the required rate of return for investments?

To begin with, the cost of capital (i.e., the required rate of return used in the net present value and annuity methods) in most companies is decided for relatively long time frames, typically a year or two. If the cost of capital changes too often, it is difficult to compare investment requests made at the beginning of the budget year with those requests made later in the year. In practice, the number and amount of investments are limited – many investments compete in the budget. Therefore, whenever possible, the different investment alternatives must have the same conditions independent of the time when they arise. To compare an investment proposal made in February with one made in November, it must be assumed that the annual investment budget is sufficient for all proposals that meet the required rate of return. This assumption implies that the budget, which is prepared at the beginning of the year, will be sufficient for all investment proposals even though they are unknown at that point.

Therefore, typically a template is used to calculate the cost of capital, especially at industrial companies in which the investments are directly related to their primary activities (e.g., new equipment, machinery, and marketing efforts). Thus, the cost of capital can vary from company to company and from case to case. Despite all this, there are still other factors that influence the size of a company's cost of capital.

In situations where the decision is whether or not to make a particular investment (e.g., when comparing this investment with other investment alternatives), the cost of capital (i.e., the required rate of return) is often determined by the alternatives. Sometimes all companies in the same industry use a common interest rate that is based on international practice.

In reality, different investments are often evaluated using different costs of capital. For example, it is possible to differentiate between *strategic investments, adaptation investments,* and *rationalization investments.* A *strategic investment* is binding on the company for a long time. Some examples are the investments required for developing a product group, building a new production plant, or opening a new market. An *adaptation investment,* which is more a tactical investment, is made to ensure that the company can sustain its operations for a relatively long time. Some examples are the investments required for investing in a new manufacturing process or in a new machine park. A *rationalization investment* is operational and is made to improve normal operations. Some examples are the investments required for developing new work methods and routines or eliminating bottlenecks in production. Typically, a lower cost of capital is used for a strategic investment than for an adaptation investment. Similarly, typically the cost of capital for a rationalization investment is set still higher. The forest industry example that follows explains this conclusion.

EXAMPLE

A small investment in an existing pulp and paper plant (that is, a rationalization investment) can be very profitable. If, for example, the company can increase its operational capacity throughout the entire manufacturing facility by eliminating a bottleneck in operations, then the investment will most certainly be very profitable. Typically, an investment in a completely new manufacturing process cannot compete with this rationalization investment. For this reason, a higher required rate of return is used with a rationalization investment and a lower required rate of return is used for a new manufacturing process.

For an investment in forests (a strategic investment in this industry), usually the discount rate is even lower. Without the security of a supply of raw materials in the long term, the company cannot use its production facilities profitably. In addition, forests are very long-term investments for biological reasons. An investment in a large forest can almost never be measured against either a short-term rationalization investment in an existing business or an adaptation investment.

Another distinction can be made between *non-mandatory investments* and *mandatory investments*. Most examples so far have been in the first category (i.e., investments made to improve financial results). An example of a mandatory investment is one required by legislation. In such cases, a discount rate might not even be used because the company has no choice other than to make the investment.

Clearly, alternative required rate of return possibilities, such as bank interest rates or bond interest rates, affect the determination of a company's cost of capital. In the long term, a company cannot use a required rate of return for its investments that is lower than bank interest rates. If a company's required rate of return only equaled bank interest rates, what would be the point of operating a business? In a simplified explanation, we can say that general interest rates in the economy have a strong impact on a company's cost of capital. A high interest rate will drive a company's cost of capital higher, and vice versa. The higher the cost of capital – that is, the higher the required rate of return – the fewer investment alternatives will seem profitable. As a result, the investment climate deteriorates. Higher interest rates also mean that a company's financing costs become higher, which is also reflected in its choice of cost of capital.

In this discussion, of course we also have to consider the requirements of a company's investors (e.g., owners) as far as their required rate of return. A company that borrows money and successively invests these sums at a rate of return that is lower than the interest rate on the borrowed money will be unprofitable over the long term. The company's investors will be less inclined to invest more money in the company in the future. Thus, a company's required rate of return should at least equal its average cost of financing its investments and operations. However, it is quite difficult to calculate the company's average cost of capital exactly. One reason is that historic financing costs are often poor predictions of current or future financing costs.

Because most companies try to finance their activities in the long term, they are often disinclined to earmark certain funds for certain investments. This is true even when general interest rates are low at the time the company plans to borrow for most of a particular investment. The company will probably use a standardized *cost of capital* to evaluate the investment. One reason is that the company can compare different investments made at different times. Furthermore, it is very difficult, if not impossible, to calculate the cost of the share of the investment capital that will be provided internally by the company. Therefore, all financing costs, such as interest rates for loans, are assumed to be included in the company's cost of capital.

Risks and sensitivity analyses

A key factor that influences the cost of capital is the investment's *risk*. High-risk investments are evaluated using a higher cost of capital than less risky investments. Usually, the cost of capital includes a *risk premium* (i.e., it is set at a higher value than is technically necessary). In addition, inflation also influences the cost of capital because future receipts decline in value in inflationary times (compare with the time preference discussion above). A period of high inflation will cause the cost of capital to increase and the amount and number of investments to decrease. Normally, it is assumed that inflation is included in the cost of capital. However, it is important to point out that inflation is only one factor among many factors that influence the size of the cost of capital. Even if inflation were zero, companies would still use a discount rate.

Finally, companies often use the cost of capital as a managerial prioritization tool. Because of budgetary limitations, alternative investments must be compared with each other. It is quite common that a company's investment proposals spring from a rather low level in the hierarchy – from the employees who work directly in operations. However, investment decisions are typically made at a higher level, even in some cases at the Board level. A high cost of capital is thus a way to prioritize (even eliminate) investment proposals. If the cost of capital is set fairly high, then only the really profitable proposals will work their way up in the hierarchy of investment proposals.

During the investment analyses, often some form of sensitivity analysis is conducted – that is, key components of the calculation are changed. For example, changes may be made in the receipts (usually the selling price of goods and services resulting from the investment), in the payments (such as the cost of labor), or in the discount rate.

5.5 Capital investment analysis in practice

Determining the required rate of return is not the only difficulty in making investment evaluations. In practice, the evaluations are much more complex than the mere calculations suggest. How do we evaluate non-financial consequences? How do we value payments/receipts in the distant future? How long is the economic life of the new equipment we plan to procure? How does a particular investment influence the opportunities for future investments? The answers to these and other questions lie behind our investment calculations; they may even be more important than the calculations.

Non-financial consequences

As we have already discussed in the context of investment risks, it is not uncommon to be blind to investment consequences that cannot be measured monetarily (e.g., work environment improvements, human resources development, and long-term research efforts). Often there is a tendency to under-emphasize these consequences. However, the very fact that we cannot measure them in numbers means they deserve special attention. Thus, it is clear that the investment calculation is only part of the broader investment evaluation.

Evaluation of future payments and receipts

Of course, there is a fundamental difficulty in evaluating the future payments and receipts (several years in the future) related to a particular investment. The more distant these amounts, the more difficult the task. It is here that the personnel with the most experience should be involved. Although both payments and receipts are difficult to predict and measure, future receipts are especially complex. Their calculation often requires predictions about uncertain market conditions, new technologies, and so on.

In some cases, there is a relatively extensive body of data to refer to. Consulting firms, financial institutions, and international organizations, for example, are good sources of these data. Such support is often used in calculations where precision is a high priority.

Furthermore, it is often quite difficult to determine which payments and receipts are associated with a specific investment. One example is the situation when a manufacturing firm invests in a new machine that is only one step in a series of manufacturing steps. How do we know if the quality improvements in the product can be attributed to this specific machine?

Additional investments and investment limitations

In practice, it is often difficult to investigate which additional investments are required to achieve the potential of the original investment. A company that acquires a new machine intended to eliminate manufacturing bottlenecks may later discover that a bottleneck has now appeared in another place in the production system, reducing the benefits of the new machine. In other situations, problems with peripheral equipment, commissions, or employee training may be overlooked.

It is the nature of investments that they create opportunities at the same time that they limit other opportunities. For example, new limitations may pop up when a manufacturing company establishes a new assembly plant. The plant location may be in a country with special laws and regulations. Or the plant, with limited capacity, may require additional investments. And so on. As an example, the acquisition of a new paper machine, which might constitute the platform for a strategic change by a pulp and paper company, is usually a major original investment with a very long life. Thus, such an investment will limit the opportunities for making other investments. Because this is the natural result of long-term investments, it is crucial that managers investigate all aspects of investments when they compare various investment alternatives.

CHAPTER 6

Accounting

Many events in which a company interacts with its environment can be accounted for as *financial transactions*. Such events include purchases (e.g., raw materials, supplies, or services), salary payments to employees, sales of goods and services, bank deposits, loans, and so on. These are events that can be quantified and recorded in the company's recordkeeping system. The term *bookkeeping* is used to describe the daily process of recording, classifying, and valuing these events; the term *accounting* is used to describe the process of analysing, interpreting, and communicating these events. However, although the two terms are often used interchangeably, accounting takes a more analytical perspective on the results that bookkeeping produces (e.g., the annual financial statements). The bookkeeping/accounting records and financial statements must comply with the governing principles, practices, and laws.

Bookkeeping uses various rules and standards to systematically record financial transactions monetarily. There are two main purposes behind this systematic process. First, recording these events allows a company to monitor its financial situation on an up-to-date basis. Second, recording these events provides protection (as well as information) for a company's stakeholders (e.g., owners, customers, suppliers, government agencies, etc.). For example, suppliers are interested in a company's ability to pay for goods and services; potential and current creditors are interested in evaluating whether given loans will be repaid; owners are interested in the return on their investments; taxing authorities are interested in whether a company has paid its payroll taxes, VAT, and corporate income taxes.

In Sweden, according to the Swedish Bookkeeping Act, a company generally has the following obligations [translated]:

- to maintain a record of all financial transactions
- to ensure that vouchers (supporting documentation) exist for all bookkeeping entries, and that the system documents the historical record of such entries
- to maintain all accounting information with the equipment/systems needed to present this information in the required form
- to prepare a balance sheet and to end the accounting period according to required procedures.

Bookkeeping requirements

The bookkeeping requirements, among other things, are the following:

- to maintain a current record of financial transactions so that they can be presented chronologically in the Journal (also called the day book) and systematically in the General Ledger
- to ensure that written documentation supports all bookkeeping entries
- to archive these financial records properly and safely for specific times periods (these periods vary by legal jurisdiction; in Sweden, the time period is 7 years)
- to prepare the annual year-end closing of the financial records and to prepare the year-end financial statements.

6.1 Current recording of transactions

The *current recording of transactions* means recording the financial transactions that take place between two year-end closings. During one year, a company's financial transactions have to be recorded currently and systematically. This means that these transactions must be recorded in various types of accounts. For many companies, such transactions occur many times a day (e.g., bill of a delivery and subsequently a related *invoice)*. Figure 6.1 is an example of a typical invoice in Sweden.

STORVIK FINMEKANIK

Faktura

Faktura nr/Kund nr		Fakturadatum
3332	50208	2016-01-07

Leveransadress	Fakturaadress
CONSTELLATION AB	CONSTELLATION AB
Bulevägen 1	Box 134
ÖSTERBYMO	570 60 ÖSTERBYMO

Er referens	ANDERS LARSSON	Vår referens	SVEN AXELSSON
Er order		Vår order	
Lev villkor		Bet villkor	15 DAGAR NETTO
Lev sätt		Förfallodatum	2016-01-22
		Dröjsmålsränta	17,00%

Art nr	Benämning	Antal	Enhet	A pris	Rabatt	Summa
77125	Hållare till styrenhet	2 000		40,00		80 000,00

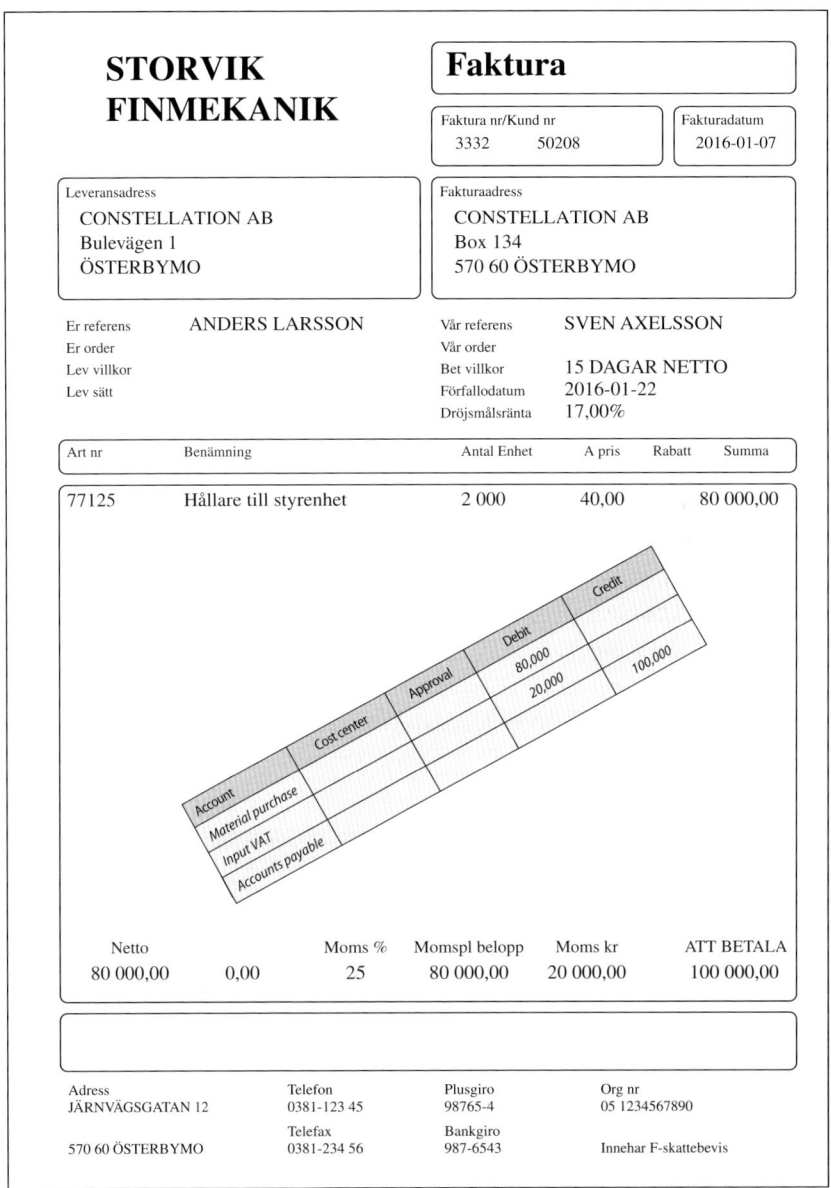

Netto		Moms %	Momspl belopp	Moms kr	ATT BETALA
80 000,00	0,00	25	80 000,00	20 000,00	100 000,00

Adress	Telefon	Plusgiro	Org nr
JÄRNVÄGSGATAN 12	0381-123 45	98765-4	05 1234567890
	Telefax	Bankgiro	
570 60 ÖSTERBYMO	0381-234 56	987-6543	Innehar F-skattebevis

Figure 6.1 A typical invoice for the purchase of materials by a mid-size Swedish company.

Invoice requirements per Swedish law

According to law, an invoice (supporting voucher) must include, among other things, the following:

- date the invoice was prepared
- customer or client
- time frame of goods/services delivered
- description of goods/services delivered (specification)
- amount of invoice.

SOURCE: SWEDISH TAX AUTHORITY

In bookkeeping terminology, an invoice received is the written, *supporting voucher*, or evidence, that supports the company's payment for a good or service. The invoice provides proof of the specific financial transaction. Other verification documents are, for example, purchase receipts, lease contracts, and bank statements (the last verifies interest revenue received/paid in a certain time frame). All such documents should be filed chronologically and numbered so they can easily be retrieved as evidence for a bookkeeping entry. In Sweden, legislation specifies how these records should be maintained.

Invoice requirements per VAT laws

The requirements for Swedish invoices for VAT are also regulated by law. According to the laws for VAT, an invoice must contain at minimum the following:

- a unique serial number
- the seller's VAT registration number
- the seller's and purchaser's name and address
- the amount and type of goods and/or services delivered
- the price of the goods/services before VAT
- the VAT base(s) that are applicable
- the VAT amount to be paid.

SOURCE: SWEDISH TAX AUTHORITY

The intention of these laws is to facilitate the tracking of all financial trans-actions so that it is possible, afterwards, to see what was done by whom, and in which connection. The underlying purpose, of course, is to protect a company's creditors and its other stakeholders, and to ensure that the financial information provided to the tax authorities is accurate.

Current accounts and verfications (Chapter 5, Swedish Bookeeping Law) [translated/summarized]

Verifications:

Paragraph 6 – Every financial transaction must have verification by a supporting invoice.

Paragraph 7 – The invoice shall contain all required information, including invoice number, details of the business event, terms, etc.

Paragraph 8 – Relative to Paragraph 7, information may be omitted (if there are difficulties), and the omission is consistent with good accounting principles.

Paragraph 9 – Revised/corrected invoices must give full and applicable details.

When an invoice is received, the invoiced amount is coded and entered in the bookkeeping system. The invoice is coded with the relevant account numbers and the amounts are allocated to each account. In the *double-entry bookkeeping system*, for every bookkeeping entry at least two accounts are involved: a *debit account* and a *credit account*. In addition, the account num-bers, the cost centers, and the invoice approval are entered.

Account	Cost center	Approval	Debit	Credit
Material purchase			80,000	
Input VAT			20,000	
Accounts payable				100,000

Figure 6.2 Coding table of the invoice received.

As an example, we can use the invoice data from Figure 6.1 to make a bookkeeping entry in the Journal, as shown in Figure 6.2 where we see a printed table with three rows and five columns. The three rows represent the financial event as shown on the supplier's invoice. Using bookkeeping terminology, we can state the following: (1) SEK 80,000 is debited to account Material purchase for the purchased materials; (2) SEK 20,000 is debited to account Input VAT for the VAT (bookkeeping for VAT is described in Section 6.5); (3) SEK 100,000 is credited to account Accounts payable (not to a cash account because at this point the invoice has not been paid). In the following section, we explain the basic bookkeeping concepts.

6.2 T-accounts and double-entry bookkeeping

A voucher, as provided by the invoice received, is recorded in three different accounts. Account Material purchase is used to record the expenditure for materials purchased. Account Input VAT is used to record the amount of VAT the supplier is required to charge (in Sweden, 25 % of the materials cost). Account Accounts payable is used to record the liability (money owed to the supplier). We will return later to the system for structuring the accounts.

Every entry in bookkeeping has two sides: a *debit* side and a *credit* side. Therefore, the descriptive term is a *T-account*, as shown in Figure 6.3.

The invoice is recorded using the double-entry bookkeeping system. This means, for every financial transaction, the total of the amounts on the debit side must equal the total of the amounts on the credit side. Compare this rule with the bookkeeping entries in Figure 6.2.

In Figure 6.2, materials and VAT are debited for a total of SEK 100,000, and the liability to the supplier is credited for a total of SEK 100,000. The reason for coding the financial transaction in this way relates to the structure

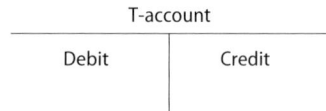

Figure 6.3 T-account.

Examples of ...	
... assets:	Intangible assets (e.g., patents), land and buildings, machinery and equipment, financial assets (e.g., leases and investments), inventories, accounts receivable from customers, cash in bank, and cash on hand.
... liabilities:	Accounts payable to suppliers, bank loans, taxes payable, and bank overdrafts.
... income:	Turnover (sales), interest, and royalties.
... expenditure:	Materials, salaries, rents, repairs, work supplies (e.g.,work clothes), office supplies, and travel expenses.

All these categories have one or more accounts. It is also worth noting there are many more expenditure accounts than income accounts.

of the General Ledger accounts. There are two principal groups of accounts: *balance sheet accounts* and *income statement accounts*. The accounts for *assets, owners' equity,* and *liabilities* are balance sheet accounts, which at the year-end closing are merged into a *Balance Carried Forward.* The accounts for *income* and *expenditures* (sometimes referred to as revenue accounts and cost accounts) are called *income statement accounts,* which at the year-end closing are summarized to a *Profit and Loss account (P&L account).*

The principal rules for the various accounts are as follows:

- *asset accounts* are increased with debits; that is, a debit to an asset account represents an increase in amount
- *liability accounts* are increased with credits; that is, a credit to a liability account represents an increase in amount
- *income* accounts are credited (except for corrections)
- *expenditure* accounts are debited (except for corrections).

There are a number of memory tricks you can use to remember these rules. One way is to learn how a particular account (e.g., Cash) works. Then you can use this knowledge to remember how other accounts work. First, Cash

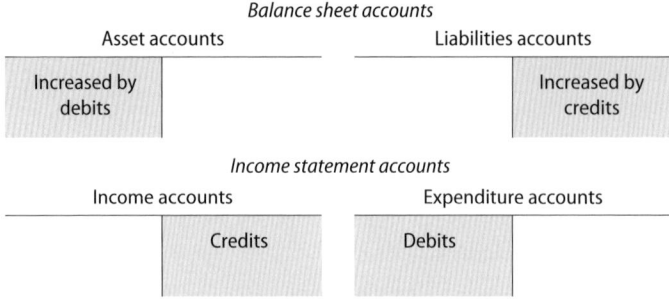

Figure 6.4 Principal accounting rules for various account categories.

is an asset account and increases with debits. We know that the rule for liabilities is the opposite (liabilities increase with credits). If a company receives income in cash we debit (increase) the Cash account and we credit (increase) the same amount to an Income account. An expenditure paid in cash works the opposite way: we credit (decrease) the Cash account and we debit (increase) the same amount to an Expenditure account. See Figure 6.4 for an illustration of the bookkeeping rules using T-accounts.

If we compare these accounting rules with our invoice example, we see they agree with how we record invoices. In the invoice example, it is clear the company has purchased materials – expenditures until the materials are consumed and sold (as cost of goods sold). The VAT is recorded as an asset (this rule is explained in Sections 6.4 and 6.5). Finally, we owe the supplier for the total sum. Figure 6.5 uses T-accounts to illustrate the transaction.

If an invoice is correct, it is paid. If the invoice appears incorrect, it must be questioned. The individual who approves the invoice also verifies that the delivery of the goods/services stated on the invoice is correct. Payment of an

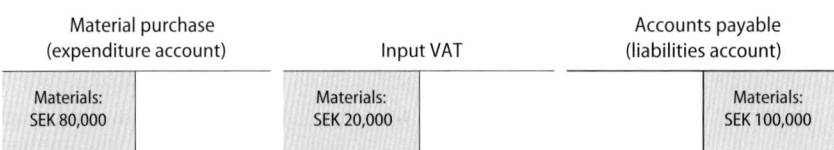

Figure 6.5 Accounting for recording the invoice.

Figure 6.6 Accounting for paying the invoice.

invoice is handled as a separate event. Typically, when a company pays an invoice from its bank account, two bookkeeping events occur. The amount in the Cash in Bank account (an asset) is reduced by the amount paid, and the Accounts Payable account (the liability) to the supplier is reduced by the same amount. Figure 6.6 illustrates the payment of the invoice from our example.

Making corrections

When a mistake is made in the bookkeeping, a correction is required. The general intention is that the correction should not eliminate the previous (incorrect entry or entries) so that it remains possible to determine afterwards who made the correction and why it was made.

A correction is made that reflects the actual financial transaction. The correction is made in two steps: (1) reversing the incorrect entry and (2) recording the correct entry. In this way, a trail of the bookkeeping events is maintained. The correction is made by simply changing the debit in the original transaction to a credit (or vice versa). Thus, the erroneous coding, so to speak, is corrected, after which you can post the transaction correctly.

Why do we use double-entry bookkeeping?

If a company uses *double-entry bookkeeping* and follows the rules outlined above for the different account categories, the result will be an elegant description of the financial events of the year and the financial position (i.e., assets and liabilities) at year-end. Two financial statements prepared at year-end are the *balance sheet* and the *income statement*. The balance sheet is a snapshot of the company's financial position at one point in time (e.g., the last day of the accounting year). The income statement presents the revenues and costs for the year. Figure 6.7 diagrams these two statements.

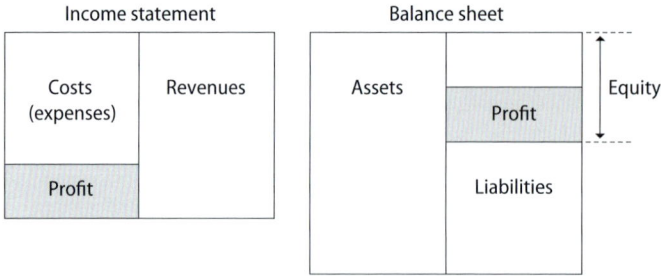

Figure 6.7 Diagram of the company's income statement and balance sheet.

Two definitions of profit (loss)

Profit (loss) = Revenues – Costs *(see the income statement)*

Profit (loss) = Change in assets – Change in liabilities *(see the balance sheet)*

Thus, a company's profit appears both on the balance sheet and on the income statement. The difference between the revenues and costs is the year's *profit* (or *loss* if costs exceed revenues). Or, from another perspective, the company's profit (or loss) is the difference between the change in assets and the change in liabilities during the year. If assets have increased more than liabilities, the company has made a profit; if liabilities have increased more than assets, the company has sustained a loss. If the rules of double-entry bookkeeping are followed, the financial transactions will be recorded properly. The financial result in both statements *must* be the same.

Closing the accounts and preparing the year's financial statements

In practice, company revenues and costs are reconciled and closed at year-end (large companies, especially publicly listed companies, also prepare quarterly or interim financial statements). This procedure is called *closing the books* or *year-end closing*. This means the company determines its total revenues and total costs for the year as well as the totals of its assets and liabilities on the *closing date* (e.g., the last day of the accounting year). At this point, the current accounts are closed.

A simple way to think about this is that the current accounts are closed against two summary accounts: the *Profit and Loss account (P&L)* and the *Balance Carried Forward (bal. c/f)*, also called the *Closing Balance*. All income statement accounts (i.e., income and expenditures) are closed to zero, and their difference is transferred to the P & L account. The totals of balance accounts (i.e., assets and liabilities) are calculated, and the difference in each account is entered in the Closing Balance.

As an example, consider how the Cash in Bank account (from above) is calculated. We must state that the single financial transaction in our example would, in reality, have been only one of many transactions. In reality, numerous transactions are recorded in the Cash in Bank account during the year. Assume, then, that the *Opening Balance* of the Cash in Bank account was SEK 1,000,000 (as the result of many previous transactions). See Figure 6.8.

Cash in Bank	
Opening balance: SEK 1,000,000	Invoice payment: SEK 100,000

Figure 6.8 The Cash in Bank account before closing.

The sum on the debit side is SEK 1,000,000 and the sum on the credit side is the payment of SEK 100,000. The difference (SEK 900,000) must be carried to the credit side of the account so that it balances. See Figure 6.9. The account is an asset account (money that the company has in the bank must naturally be an asset), and therefore it is closed to the Closing Balance account. As the Cash in Bank account is credited, so then is the Closing Balance account debited.

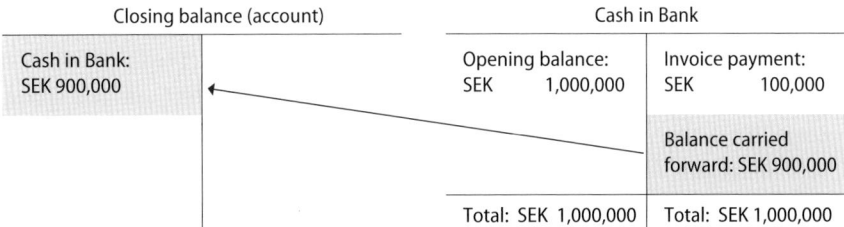

Figure 6.9 The Cash in Bank account closed to the Closing Balance account.

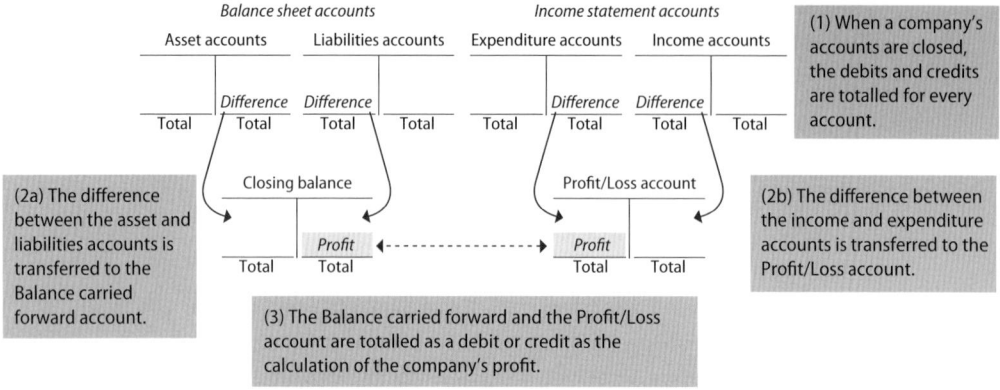

Figure 6.10 Closing of the company's accounts to the Closing Balance and to the Profit/Loss account.

All other accounts are closed in the same way, which means all income statement accounts (i.e., the income and expenditure accounts) are closed to the Profit and Loss account, and all balance sheet accounts (i.e., the asset and liability accounts) are closed to the Closing Balance account. See Figure 6.10.

Finally, the Profit and Loss account and the Closing Balance account are closed. The difference between the total debits and total credits for each account, as the result of the double-entry bookkeeping, must equal. If the difference is a credit for the Closing Balance account, the company has made a profit for the year; if the opposite is true, the company has sustained a loss for the year. The number that balances (total debits = total credits) for both accounts is the year's profit or loss. In our extended example, there is a profit of SEK 843,000. See Figure 6.11.

From the presentation shown in Figure 6.11 of the company's assets and liabilities and its revenues and costs, we recognize the principal features of the balance sheet and the income statement that were presented previously in the chapter. See Figure 6.7. The difference between the debits and the credits constitutes the year's profit. We can see that the year-end balances of both sets of accounts are equal (debits = credits), and thus that the bookkeeping follows the principles of double-entry bookkeeping.

Closing Balance account

Assets	SEK	Liabilities & Equity	SEK
Buildings	3,025,000	Share capital	600,000
Machinery	2,312,000	Balance carried forward	3,944,000
Inventory	2,017,000	Profit for the year	843,000
Cash in Bank	900,000	Bank loans	4,000,000
Cash	347,000	Customer advances	400,000
		Accounts receivable	2,011,000
		VAT	373,000
Total:	12,171,000	Total:	12,171,000

Profit/Loss account

Costs	SEK	Revenue	SEK
Purchase of goods and material	3,736,000	Net sales	9,080,000
Facility costs	17,000	Other operational revenues	17,000
Energy cost	480,000	Interest revenues	47,000
Office consumables	43,000		
Telephone and postage	125,000		
Salaries	3,082,000		
Depreciation	678,000		
Other operational costs	40,000		
Interest costs	295,000		
Profit:	843,000		
Total:	9,339,000	Total:	9,339,000

Figure 6.11 Closing Balance and Profit/Loss accounts.

In practice, neither the Profit and Loss account nor the Closing Balance account is used today. Instead, one goes directly to the preparation of the income statement and the balance sheet (if bookkeeping software is used, this procedure is automatic). However, thinking about the Profit and Loss and the Closing Balance is a simple way to understand how the income statement and the balance sheet are prepared. See Figure 6.11.

The year-end financial statements (annual accounts) are presented in the company's *annual financial statements* according to the Bookkeeping Act (in Sweden). The accounts are listed in a particular order. The design of the financial statements, and related rules, are described in more detail in Chapter 7.

6.3 Systematic order of the accounts

As described above, usually a company's accounts are classified in four groups: assets, liabilities, income, and expenditures. This classification is further refined to create a detailed structure for revenues and costs. Most Swedish companies use a standardized *chart of accounts* called *BAS 2015* that is updated annually (see http://www.bas.se). In the example in this chapter, we use account names/classifications from this chart of accounts. The advantage of applying the chart of accounts is that it lists standard account names, numbers, and classifications. With minor adjustments, most businesses, associations, and government agencies in Sweden can use the BAS chart of accounts, and most software accounting programs in Sweden support the BAS chart of accounts. In addition, most textbooks use this account structure as their starting point.

By the use of such a standardized chart of accounts, a Swedish company meets the legal requirements for the external accounts review – the annual *audit*, which verifies the fairness of the internal bookkeeping. In addition to this external audit, the use of the standardized chart of accounts facilitates the internal use of the accounts as data sources for cost calculations and other analyses. However, there are no legal requirements for such internal usages.

Systematically and chronologically

According to the Swedish Bookkeeping Act, all financial transactions should be accounted for systematically and chronologically. Use of the *Journal* (in which transactions are recorded in a sequence according to time of occurrence) meets the chronological requirement. Use of the *General Ledger* (in which transactions are sorted by account) meets the systematic requirement. The year-end results, when the balance sheet and income statement are prepared, are recorded in the *Annual Accounts Book*. See Figure 6.12.

In very simple accounting systems, the Journal and the General Ledger are not separate records. For example, a cashbook is used (often referred to as the column diary) in which each column is an account and each row a transaction. See Figure 6.13. The column diary satisfies both the systematic and chronological requirements.

138

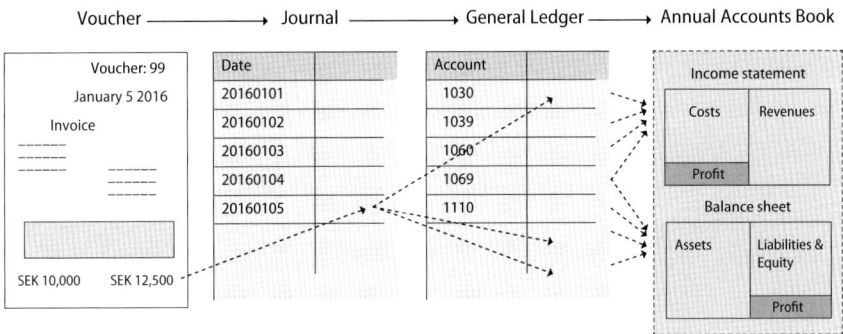

Figure 6.12 Principal formats for double-entry bookkeeping.

Date	Cash (debit)	Cash (credit)	Cash in Bank (debit)	Cash in Bank (credit)	Sales (credit)	Material (debit)	Salaries (debit)	Other (debit)
January 7 2016				100		100		
January 10 2016	7,200				7,200			
January 15 2016		1,100					1,100	

Figure 6.13 Column diary.

Today, nearly every company, even small companies, uses a bookkeeping software program that simplifies the arithmetic calculations of bookkeeping. In a computer-based bookkeeping system, every invoice is a separate entry. These records have approximately the same information fields as an accounting table. See Figure 6.2. See also Figure 6.14 for an example of an entry in a simple bookkeeping system.

A computer-based bookkeeping system can sort the financial transactions both chronologically and systematically (i.e., as the Journal and General Ledger do). Such programs can often perform other functions, for example, year-end closing, preparation of financial statements, and VAT calculations. They can also link to the company's other financial systems.

Transaction date	Recording date
January 7 2016	January 7 2016

Specification

Purchase of materials

Line	Account	Description	Debit	Credit
1	2440	Accounts payable		100,000.00
2	4010	Material purchase	80,000.00	
3	2641	Input VAT	20,000.00	
4				

Difference: 0.00   Total: SEK 100,000.00   SEK 100,000.00

Figure 6.14 Bookkeeping entry.

6.4 Bookkeeping for VAT

VAT (value added tax) is a government tax on the value added that goods and services provide to the purchaser. Businesses in Sweden and in most EU countries are required to add VAT (a nationally mandated percentage of the sales price) when they sell goods/services. In turn, business companies can deduct the VAT from the total cost of goods/services they purchase. Thus, each business company pays VAT to the government only on the difference between these two amounts. Therefore, VAT is assessed at every step in the value-creating process. In Sweden, the VAT percentages vary with the type of goods or services. Three percentages are used (2015): 6 %, 12 %, and 25 % of the basic sales price.

Four separate accounts are used for recording VAT in BAS 2015, three accounts for Output VAT (one for each percentage rate) and one for Input VAT.

The VAT paid for a purchased product is called the purchaser's *Input VAT*. As in our example, the Input VAT is recorded as a negative liability (i.e., on the debit side of a liability account). In other words, Input VAT is, in a sense, a kind of asset that the Tax Authority is required to refund. When a product is sold, the VAT paid to the seller by the purchaser is called *Output VAT*, which is recorded as a liability to the Tax Authority.

> **VAT bookkeeping**
>
> A company that buys goods/services (without immediate cash payment) records the following:
>
> - a debit to an expenditure account (e.g., Material purchase)
> - a debit to a VAT account (Input VAT)
> - a credit to a balance sheet account (e.g., Accounts Payable – trade)
>
> A company that sells goods/services (without immediate cash receipt) records the following:
>
> - a credit to an income statement account (e.g., Sale of goods)
> - a credit to a VAT account (Output VAT)
> - a debit to a balance sheet account (e.g., Accounts Receivable – trade).

If a company has more Output VAT than Input VAT, it pays this difference to the tax authorities. If, instead, Input VAT exceeds Output VAT, then the company may deduct this amount in its tax declaration. In other words, Input VAT is deductible for businesses. Of course, this rule does not apply for expenses that are not deductible, for example, certain business entertainment expenses. Moreover, private individuals (the final consumers), who pay VAT for the goods/services they purchase, but do not sell anything, are always the final payers of VAT in society.

6.5 Accurate bookkeeping is not easy in practice

In theory, bookkeeping is a rather simple system for accounting for the money associated with a company's financial transactions. By following the bookkeeping rules, a company can create a comprehensive structure for recording all kinds of revenues, costs, assets, and liabilities. However, the comprehensiveness of this structure in practice creates the bookkeeping difficulties.

Often someone in the company wants to know how much a particular resource costs. Or someone needs information for product costing or investment decisions. The source of such information is generally the bookkeeping records, where all the revenues and costs have been recorded. For example, when recording an invoice, someone must decide which General Ledger accounts to use. This decision means that the way the actual activity of recording is done has a significant impact on how various costs are officially reported. Typically, many individuals are involved in the different steps of the bookkeeping process. Therefore, there is ample room for misunderstandings and misinterpretations in classifying financial transactions and in the choice of accounts.

6.6 Management accounting

Companies often require information that the financial accounting does not provide because it is directed primarily to external stakeholders. An example of such information is the following: how many products did the company produce during a specific period, and at what cost? If a company has developed routines for tracking such information, then the company can evaluate the efficiency of its manufacturing operations. To answer such a question, two things are required: first, the company must keep a record of its production output; second, the company must keep a record of the resources used in the production processes (raw materials, labor, etc.). To determine if a system is efficient, documents must be available that provide the required data.

In complex operations, it is not enough to monitor the operations simply by recording the external financial transactions (i.e., the transactions between the company and its external stakeholders). Information on other internal activities is required. Therefore, most major companies keep their own internal accounting records (e.g., different types of *budgets*) for control of the operations on a continuous basis. These internal accounting records are referred to as the company's *management accounting* system (to be distinguished from its financial accounting system described above). Management accounting is thus used to provide company managers with financial and other information used for management control.

Management accounting
– a reflection of the company's operations

The aim of management accounting is to express the company's operations processes in monetary terms. For example, the management accounting system provides the following information:

- cost of the company's goods/services
- profitability of the individual goods/services, of product groups, of markets, etc.
- the efficiency of different departments and units
- the appropriateness of the overhead costs allocations to goods/ services (see Chapter 4).

To acquire this information from, for instance, the manufacturing processes, a company must track the movement of raw materials inventories to the manufacturing and assembly (work in progress), to completion (the inventory of finished goods), and, finally, to the sale to customers. See Figure 6.15.

However, to evaluate the efficiency of a department or unit in a certain period, we need a *performance measurement*, which accounts for the amount of raw materials consumed, the number of machine and labor hours used, etc.

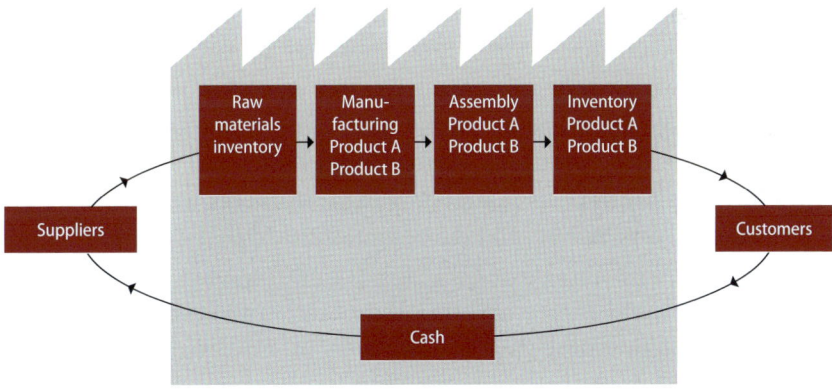

Figure 6.15 Materials processing at a manufacturing company.

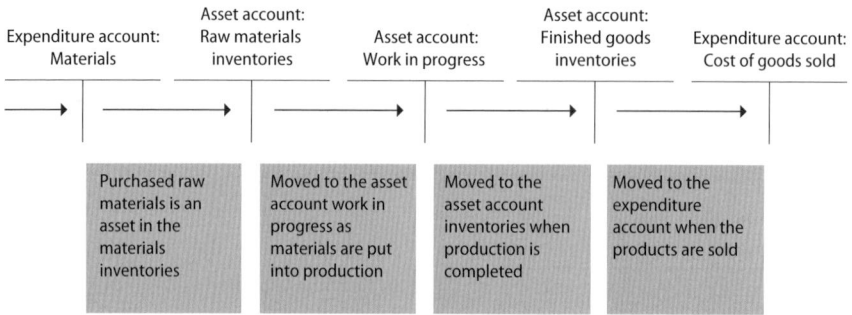

Figure 6.16 The value of resource consumption, as exemplified by the movement of materials in connection with the manufacturing process.

Financial accounting records the company's financial transactions with external actors (e.g., recording an incoming invoice). In the management accounting system, other kinds of events are recorded (e.g., movement of raw materials to the factory floor). In this event, the materials are not consumed but are put into production; the materials are now called *work in progress.* When production is finished, the materials have become finished goods inventory.

Figure 6.16 illustrates the flow of materials from the raw materials inventories to the finished goods inventories. This flow, as well as the use of other manufacturing resources, must be recorded in the management accounting system. These other resources include, for example, direct labor and machinery/utilities usage. These resources are transferred between various accounts in the management accounting system until they end in the account for finished goods inventories. When the finished goods are sold, the cost is recognized (cost of goods sold).

Companies use various manufacturing documents to track this flow, such as the bill of materials and the operations card that itemizes the time each activity takes. This means that the accounting does not necessarily match, for example, the actual time an operation took compared to the planned completion time (that is, there may be a difference between the actual time and the estimated time). Thus, the accounting records a predetermined amount for the use of labor, machine time, etc. based on the number of units produced.

CHAPTER 7

Financial accounting
– reporting to external stakeholders

Publicly owned companies are required by law to prepare an *annual report,* also called *annual financial statements,* at year-end. The annual report typically consists of an *income statement,* a *balance sheet,* and an *administration report (management report).* Together, these documents report on the company's annual results and its year-end financial position. Attached to the annual report is the independent *auditor's report* that reports on the external examination of the company's annual report and on company management by the Board of Directors and the Managing Director (also referred to as the CEO, Chief Executive Officer).

The purpose of a Swedish company's annual report, which is a public document that is registered with the Swedish Companies Registration Office (in Swedish: *Bolagsverket*) is to provide the various external stakeholders with insights into the company's operations. Based on the annual report, it should be possible to determine if a company is profitable, if it can pay its bills, if it can repay its lenders, and so on. This chapter describes how companies prepare the financial statements in their annual report – income statement and balance sheet – from their accounts. The chapter also discusses how to interpret these financial statements.

7.1 Financial accounting

Financial accounting allows stakeholders not directly involved in a company's operations to evaluate the company. This evaluation is based, in part, on the company's results for a specific time period and on the company's financial position at the end of that time period. The evaluation also allows these stakeholders to make sure that financial or other resources have not

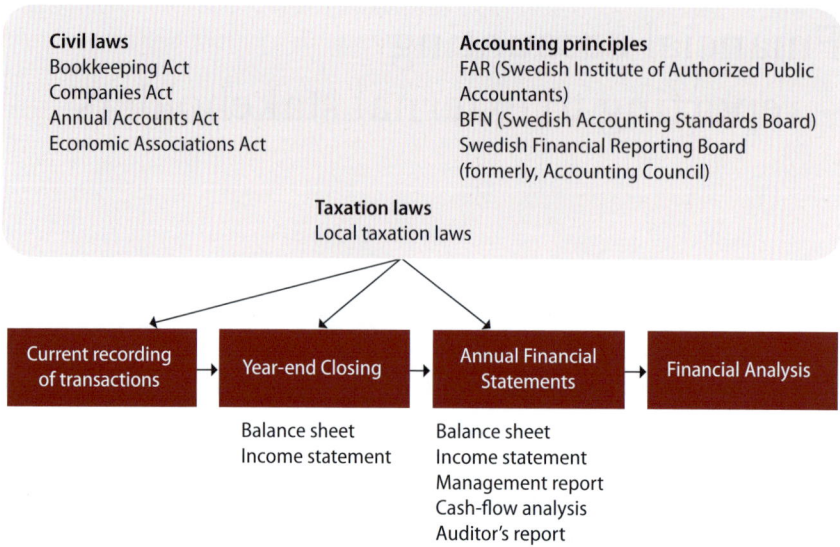

Figure 7.1 Laws governing companies' financial accounting.

been improperly removed from the company. To facilitate this evaluation, the independent auditor examines the accounts and the financial statements. After this examination, the independent auditor issues a report on the fair presentation (according to generally accepted accounting principles) of this information.

Thus, *financial accounting* is mainly directed at external stakeholders such as shareholders, suppliers, customers, financiers, governmental authorities (e.g., taxation authorities), and the general public. See Figure 7.1. The boundary between management accounting and financial accounting, however, is fluid. The *current recording of transactions* provides input for both. Furthermore, financial accounting is often used in management accounting as well. This situation is particularly common at smaller companies that sometimes only maintain financial accounts.

Unlike management accounting, financial accounting is regulated by law. These laws are, in Sweden, the *Annual Accounts Act,* the *Bookkeeping Act,* the *Municipal Tax Act,* the *Companies Act,* and the *Economic Associations Act.* These laws govern financial accounting, bookkeeping, taxation, and so on. Moreover, actual practice, developed continuously over time, for these activities is also important. This is called "good accounting practice" as determined by the professional association *Swedish Institute of Authorized Public Accountants (FAR)*. Recommendations from, among others, the *Swedish Accounting Standards Board* (BFN) and the *Swedish Financial Reporting Board* are also important.

A fundamental principle behind these laws, principles, and practices is that society is based on the economics of business and trade. Therefore, it is essential that accounting/auditing laws, principles, and practices support these activities. For the benefit of all society, people must be able to trust one another. Society as a whole suffers when lack of trust undermines people's willingness to invest in companies, deliver goods and services, and lend money.

Civil law and *tax law*, to some extent, have conflicting objectives with respect to financial accounting. See Figure 7.2. The main objective of civil law (for example the Annual Accounts Act, the Companies Act, and the Economic Associations Act) is to protect owners, lenders, suppliers, customers, and other stakeholders from the possibility that a company will over-state its earnings and financial position. The objective of the tax law (primarily the Municipality Tax Law) is to protect society from the possibility that a company will under-state its earnings and thus under-pay its taxes. See Section 7.4.

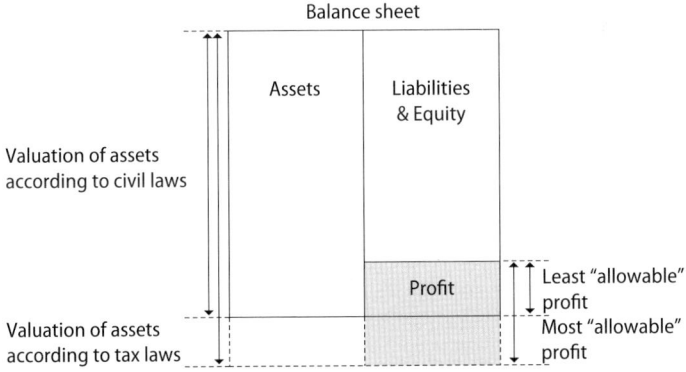

Figure 7.2 Civil law and tax law have different objectives. In the diagram, for illustrative purposes, the company's profit appears farthest down on the liability side of the balance sheet.

Limited liability companies

The general terms "company" or "business" are not legal concepts. A business can be organized in different legal forms. Perhaps the most important legal form is the *limited liability company*. In principle, all large companies (in Sweden and elsewhere) are limited liability companies but many small companies in Sweden use this legal form as well. In fact, the majority of all limited liabilities companies in Sweden are small businesses (with ten or fewer shareholders). Single-owner limited liability companies are also common. There are several other legal forms for businesses. Besides the limited liability companies, there are the following:

- condominiums (tenant-owners' associations)
- economic associations (co-operatives)
- sole proprietorships
- partnerships
- limited partnerships
- not-for-profit associations
- joint ownership associations
- foundations (trusts).

After the sole proprietorship, the limited liability company is the most common form of business. The limited liability company is a legal entity that combines the ability to raise capital (financing) with exemption from personal liability for the shareholders. In other words, it is a corporate structure whereby the owners of the company cannot be held personally liable for the company's debts (liabilities). Thus, shareholders in a limited liability company limit their risk to the amount of capital they have invested. The most distinguishing feature of the limited liability company is, as its name suggests, limited liability.

There are extensive laws that govern limited liability companies. In Sweden, the Companies Act is the primary source of these laws, complemented by the Bookkeeping Act and the Securities Market Act. Many of these rules are legally binding and cannot be negotiated. A limited liability company must always use the words "limited liability" in its name (in English, abbreviated as Ltd.; in Swedish: *aktiebolag*, abbreviated as AB).

The company must also be registered with the Swedish Companies Registration Office. After registration, the limited liability company becomes a legal entity.

The financial foundation of the limited liability company is the share capital from the shareholders. Because the shareholders take no personal responsibility for the company's liabilities, this share capital is to protect the company's *creditors*, that is, the actors to whom the company owes money. Assets corresponding to the share capital shall always exist within the company and cannot be distributed to the shareholders. In short, the share capital is the price the shareholders pay for the protection of limited liability. The minimum share capital in Sweden today (year 2016) is SEK 50,000, which should be structured into shares of equal amount, the *quota value* (previously, nominal value). The sum of the shares' quota value has to be equal to the share capital. A company with SEK 50,000 in share capital may have 500 shares at SEK 100 each, or 250 shares at SEK 200 each, and so on. However, the quota value is not the same as each share's market value, which is the price a purchaser is willing to pay for each share.

By law, a limited liability company must have an *Annual General Meeting* (AGM), assembling the company's shareholders. It also must have a Board of Directors and an external auditor. All limited liability companies in Sweden must also have a Managing Director (MD). The AGM is the company's highest decision-making body. All shareholders in a company have the right to attend the AGM and to vote on various proposals, based proportionally on the number of shares owned. In small and medium-size companies, generally only a few people own the shares. Larger companies often have many, widely dispersed shareholders, including financial and other institutions. Very large companies may have thousands of shareholders. In reality, however, generally a few actors own large blocks of shares and control these large companies. At the AGM, the shareholders elect the Board of Directors, who are ultimately responsible for the company's operations, and an external auditor, who audits the operations on behalf of the owners.

A limited liability company is a taxable entity, which means it pays taxes without reference to how its shareholders are taxed. For instance, the company pays taxes on its annual profit. If the profit is distributed as dividends to the shareholders, they pay tax on these dividends (this system results in the double taxation of company profits).

7.2 The annual financial statements – balance sheet and income statement

The first step in the preparation of the annual financial statements is to pre-pare the annual report, which consists of two main statements: the *balance sheet* and the *income statement*. This step can be illustrated using T-accounts. See Figure 7.3.

One simple way to understand these statements is to draw an analogy between the company's assets (i.e., cash on hand, cash in bank, accounts receivable, inventories, machinery, buildings, etc.) and the water in a basin. See Figure 7.4. The income statement shows the inflow of water (revenues) and the outflow of water (costs) during the year. The balance sheet shows the water level in the basin (value of equity) at the end of the year. Profit can be calculated in two ways: the difference in the basin's inflow (revenues) and outflow (costs) during the year, or the difference in the basin's water level (value of equity) from the beginning of the year to the end of the year.

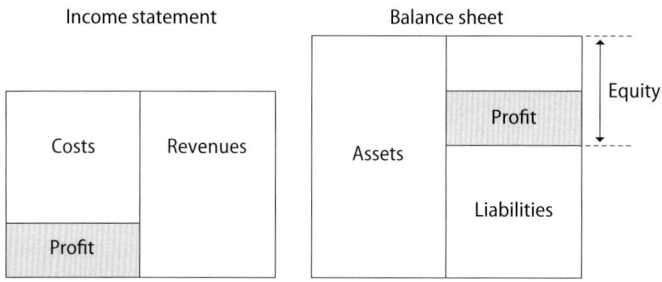

Figure 7.3 Diagrams of income statement and balance sheet illustrated as T-accounts.

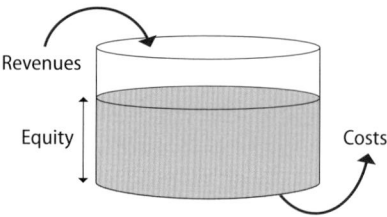

Figure 7.4 Revenues and costs illustrated as cash flows, in and out, of the company.

The balance sheet – a review

The balance sheet has two sides: *the asset side* and *the equity and liability side*. The total of assets equals the total of the liabilities (debts) and equity. Figure 7.5 presents a simplified version of a balance sheet. Figure 7.7 presents a complete balance sheet following the format from *FAR Swedish Institute of Authorized Public Accountants*.

Assets are classified under two main headings:

1. fixed assets
2. current assets.

Fixed assets include assets that are used and wear out over long time periods. Their purchase expenditures are allocated over several years through the *depreciation process*. Examples of such assets are vehicles, machinery, equipment, buildings, patents, etc.

Assets	Liabilities & Equity
Fixed assets	**Equity**
Intangible assets • Patents, goodwill, etc. Tangible assets • Land and buildings • Machinery and equipment Financial assets • Shares in other companies • Deferred tax asset • Long term receivables	Share capital Funds Profit/loss brought forward Annual profit/loss
	Untaxed reserves Accumulated additional depreciation Tax allocation reserve
Current assets	**Non-current liabilities**
Inventories etc. • Raw materials and consumables • Products in progress • Finished products and goods for sale Current receivables • Accounts receivable – trade • Other receivables • Prepaid expenses and accrued income Cash and bank balances	
	Current liabilities Advance payments from customers Accounts payable – trade Current tax liabilities Accrued expenses and deferred income

Figure 7.5 Simplified presentation of the balance sheet.

Current assets include assets that are consumed (or replaced) within one year. Examples of such assets are cash on hand, bank balances, accounts receivable, raw materials inventory, products in process, finished goods inventory, etc.

The liabilities and equity side is structured under four main headings:

1. equity
2. untaxed reserves
3. non-current liabilities (long-term debts)
4. current liabilities.

Equity is the part of the company's capital that belongs to its shareholders or owners, that is, the people who have invested in the company. This investment has (hopefully) increased over the years as the result of company profits. In other words, company equity consists of *share capital, annual profit,* and *profits brought forward* (i.e., profits from previous years that have not been distributed to shareholders).

Untaxed reserves are a company's profits that have not yet been taxed. See Section 7.4 for a discussion of how untaxed reserves are handled.

Non-current liabilities are a company's long-term debts, which means loans with a maturity date of more than one year. These debts are often loans from banks and other credit-issuing institutions.

Current liabilities are a company's debts with a maturity date of one year or less than one year. Examples of these liabilities are *accounts payable – trade (to suppliers), advance payment from customers,* and *tax liabilities.*

The Annual Accounts Act regulates the presentation of the *balance sheet.* The balance sheet in Figure 7.7 follows this presentation. The assets are listed in reverse order of liquidity. Thus, the assets that are most difficult to convert to liquid assets are listed first, and the assets that are readily available (i.e., cash on hand and bank balances) are listed last. The order for equity and liabilities is the same as the order for assets. Equity is listed first, followed by the untaxed reserves (in part equity and in part deferred tax liability). Next are the non-current liabilities and finally the current liabilities. Assets, on the left side of the balance sheet, are resources that are used in the company operations and have value; the liabilities and equity, on the right side of the balance sheet, finance these assets. In Sweden, the asset side (left) is referred to as the active side; the liability side (right) is referred to as the passive side.

The *annual profit* appears on the liability (passive) side of the balance sheet as a part of equity. The explanation is that the profit as well as the rest of the equity belongs to the owners. Thus, the company has a kind of debt or liability to its owners.

The amount of liquid funds (cash) a company has is, of course, unrelated to the amount of its profit. If we assume that a company's liabilities and equity are the same at the beginning of the year as at the end of the year, but that the assets have increased by, for example, SEK 10 million, then the company's profit for the year is SEK 10 million. However, a profit of SEK 10 million tells us nothing about how the company's cash changed or was used during the year. The SEK 10 million may be anywhere in the company's assets: cash on hand, cash in bank, inventories, machinery, etc.

By *annual profit*, we refer to taxed profit. A company is allowed to allocate a portion of its profits (before tax) to the *untaxed reserves*. These untaxed reserves are a hybrid between liabilities and equity because, in part, they consist of a deferred tax liability. If necessary, a company can use these untaxed reserves to offset future losses and thereby equalize the tax burden between profitable and unprofitable years.

There are two types of equity: restricted equity and non-restricted equity. *Restricted equity* includes, among other things, the share capital and associated funds, while the *non-restricted equity* (disposable equity), among other things, includes *profit or loss brought forward* (i.e., the accumulated profits/losses from previous years) and the *profit (loss) of the year* in question. Successful limited liability companies often distribute dividends to their shareholders every year. Dividends to shareholders are only allowed from the non-restricted equity. The Board of Directors proposes the dividends at the Annual General Meeting (AGM), which makes the formal decision.

The income statement – a review

The income statement, as mandated by the *Annual Accounts Act,* presents the company's revenues and costs during the year. Figure 7.6 is a simplified presentation of an income statement. Figure 7.8 presents a complete income statement according to the format from FAR Swedish Institute of Authorized Public Accountants.

The income statement is usually presented in report form, unlike the balance sheet, which is usually presented in account form, and revenues and costs are listed in a particular order. The Annual Accounts Act approves two variations of the income statement as far as the presentation of the costs: classified by the *function of expense,* or classified by the *nature of expense.*

The difference lies primarily in how the company divides its accounts for operating costs into more detailed sub-accounts. These sub-accounts are listed either on the basis of the company's various *functions* (production, sales, administration, etc.) or on the basis of the company's classification of its *costs* (raw materials, labor, etc.). In the presentation below, we follow Figure 7.6, which has one heading for all operating costs. Contrast Figure 7.6 with Figure 7.8 that presents an income statement classified by the function of the expense, and an income statement classified by the nature of the expense).

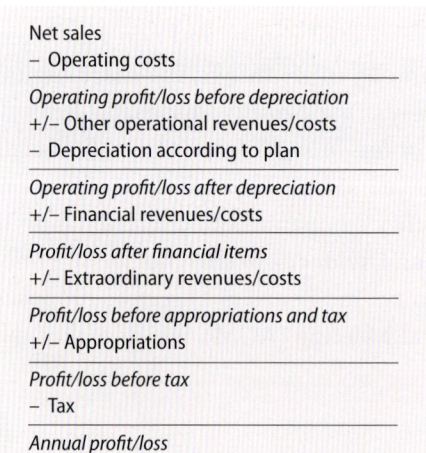

Net sales
– Operating costs

Operating profit/loss before depreciation
+/– Other operational revenues/costs
– Depreciation according to plan

Operating profit/loss after depreciation
+/– Financial revenues/costs

Profit/loss after financial items
+/– Extraordinary revenues/costs

Profit/loss before appropriations and tax
+/– Appropriations

Profit/loss before tax
– Tax

Annual profit/loss

Figure 7.6
Simplified presentation of the income statement.

Two abbreviations

EBITDA (Earnings before Interest, Taxes, Depreciation, and Amortization) =
= Operating profit before depreciation and amortization

EBIT (Earnings before Interest and Taxes) =
= Operating profit after depreciation and amortization

Net sales, also called *net turnover* (i.e., sales excluding VAT) is the first item in the income statement. This is the sum of the company's revenues from the sales of goods and services in its normal business operations. Normal business operations, of course, may be anything from Internet-based services (music and movie subscriptions, payment services, etc.), manufacturing (cars, refrigerators, etc.), consulting services (investigative tasks, management advice, etc.) to financial services (asset management, etc.).

Operating costs (expenses) are the costs incurred in the company's normal business operations. For example, operating costs include salaries, supplies, rents, and utilities.

Other operating revenues and costs are incurred in the business operations, but these revenues or costs are not directly part of the normal operations. For example, when a company sells an obsolete machine (a fairly common occurrence), the revenue from this sale is not part of normal operations. The company's business is selling products manufactured by the machine, not selling machines. An example of *other operating costs* is the cost associated with scrapping a machine that has no market value.

Depreciation according to plan is the annual decrease in value of the company's fixed assets (long-term assets such as vehicles, machinery, equipment, and buildings). In general, the company acquires most such assets with the intention of using them for several of years. Therefore, to match the costs of these assets to the revenue they generate (over many years), the company depreciates (writes-off, in ordinary usage) these assets over their expected economic life. For example, if the economic life of a machine is 10 years, we could annually depreciate 10 % of the machine's original expenditure over a period of 10 years. The assumption is that this procedure reflects the consumption of the asset. For certain intangible assets (e.g., patents), *amortization* is the English term for this kind of annual expensing.

Profit/loss from financial items, which, for example, includes interest on bank balances, is not generated from normal business operations. Industrial companies, for example, are not primarily in the financial investment business, but they can still generate substantial financial revenues. Similarly, *financial costs* arise when the company borrows money from banks and other financial actors.

Extraordinary revenue arises from unusual and non-recurring events. Industrial companies, for example, may have such revenue from the sale of a subsidiary or the sale of a license (such as the right to use a patent). These events occur rather rarely. *Extraordinary costs* are equally rare and typically are the result of isolated events.

The last items in the income statement are the *appropriations* and *taxes*. Under the Swedish tax law, a company may exclude (or rather, postpone) a portion of its profits from taxation by the creation of the *untaxed reserves* through appropriations. A company can choose between two appropriation methods: (1) *additional depreciation* (more depreciation than planned depreciation) or (2) *tax allocation reserve* (in Swedish: *periodiseringsfond*). See Section 7.4. Appropriations are essentially methods for moving some of the company's profit from the *equity* to the *untaxed reserves* on the balance sheet. This procedure is as follows: Debit a cost account, which reduces profit (which is part of equity), and credit a liability account in the amount of the untaxed reserves.

ASSETS

Subscribed capital unpaid

Fixed assets

Intangible assets
Capitalized expenditure for research and development and similar
Concessions, patents, licenses, trademarks and similar rights
Rights of tenancy and similar rights
Goodwill
Advance payments for intangible assets

Tangible assets
Land and buildings
Plant and machinery
Equipment, tools, fixtures and fittings
Construction in progress and advance payments for property, plant and equipment

Financial assets
Participations in group companies
Receivables from group companies
Participations in associated companies
Receivables from associated companies
Other securities held as non-current assets
Loans to partners or related parties
Deferred tax asset
Other long-term receivables

Current assets

Inventories etc.
Raw materials and consumables
Produts in progress
Finished products and goods for resale
Work in progress
Advance payments to suppliers

Current receivables
Account receivable – trade
Receivables from group companies
Receivables from associated companies
Other receivables
Prepaid expenses and accrued income

Investments in securities etc.
Participation in group companies
Other investments in securities etc.

Cash and bank balances

EQUITY AND LIABILITIES

Equity
Restricted equity
Share capital
Revaluation reserve
Statutory reserve

Non-restricted equity
Share premium reserve
Profit or loss brought forward
Profit (loss) for the year

Untaxed reserves

Provisions
Provisions for pensions and similar obligations
Deferred tax liability
Other provisions

Non-current liabilities
Bond loans
Liabilities to credit institutions
Liabilities to group companies
Liabilities to associated companies
Other liabilities

Current liabilities
Liabilities to credit institutions
Advance payments from customers (may also be accounted for as a deduction from Inventories etc.)
Account payable – trade
Bills payable
Liabilities to group companies
Liabilities to associated companies
Current tax liability
Other liabilities
Accrued expenses and deferred income

MEMORANDUM ITEMS

Pledged assets and contingent liabilities
Pledges and equivalent collateral to secure own liabilities and commitments reported as provisions, classified by type
Other pledges and equivalent collateral, classified by type
Contingent liabilities
Pension obligations not recognized as liabilities or provisions and not covered by pension fund assets
Other contingent liabilities

Figure 7.7 Balance sheet according to the recommended format of *FAR Swedish Institute of Authorized Public Accountants.*

INCOME STATEMENT FOR ... (year)

FAR's format for the income statement classified by nature of expense, in accordance with RedR1*

Net sales
Change in inventories of products in progress, finished goods and work in progress
Work performed by the company for its own use and capitalized
Other operating income

Operating expenses:
Raw materials and consumables
Goods for resale
Other external costs
Employee benefit expenses
Depreciation/amortization and impairment (as well as reversals) of property, plant and equipment and intangible assets
Impairment of current assets in excess of normal impairment
Other operating expenses
Operating profit (loss)

Profit (loss) from financial items:
Profit (loss) from participations in group companies
Profit (loss) from participations in associated companies
Profit (loss) from other securities and receivables accounted for as non-current assets (with separate disclosure of income from group companies)
Other interest income and similar profit (loss) items (with separate disclosure of income from group companies)
Interest expense and similar profit (loss) items (with separate disclosure of expenses referring to group companies)
Profit (loss) after financial items

Extraordinary income
Extraordinary expenses
Appropriations
Tax on profit for the year (income tax, current and deferred)
Other taxes
Net profit (loss) for the year

INCOME STATEMENT FOR ... (year)

FAR's format for the income statement classified by function of expense, in accordance with RedR1*

Net sales
Cost of goods sold
Gross profit (loss)

Selling expenses
Administrative expenses
Research and development costs
Other operating income
Other operating expenses
Operating profit (loss)

Profit (loss) from financial items
Profit (loss) from participations in group companies
Profit (loss) from participations in associated companies
Profit (loss) from other securities and receivables accounted for as non-current assets (with separate disclosure of income from group companies)
Other interest income and similar profit (loss) items (with separate disclosure of income from group companies)
Interest expense and similar profit (loss) items (with separate disclosure of expenses referring to group companies)
Profit (loss) after financial items

Extraordinary income
Extraordinary expenses
Appropriations
Tax on profit for the year (income tax, current and deferred)
Other taxes
Net profit (loss) for the year

* RedR1 is a set of recommendations for good accounting practices published by FAR.

Figure 7.8 Income statement according to the two recommended formats of *FAR Swedish Institute of Authorized Public Accountants.*

7.3 Administration report, cash flow analysis, interim report, and the auditor's report

In addition to the two main financial statements (the balance sheet and the income statement), the annual report includes an *administration report (management report)* prepared by the Board of Directors. Larger companies are also required to include a *cash flow analysis*. In the *auditor's report*, which accompanies the financial statements in the annual report, the independent auditor confirms the examination of the financial statements as well as presents an opinion on the fairness of these statements in accordance with good accounting practice and the law. Most major share companies also prepare and publish reports (*quarterly reports*) on a quarterly basis.

The *administration report* describes the company's activities and results in the past year and comments on its future plans. According to the Swedish Companies Act, the administration report is required to present a fair review of the company's operations, financial position, and results. For example, the following information may appear in the administration report:

- events and conditions not recognized in the financial statements but which are nevertheless of importance in assessing the company's financial position
- important events during the financial year, or after the end of the financial year
- a description of the company's research and development activities.

The administration report of limited liability companies, and some other business entities, should also include proposals for how the year's profit or loss should be handled. Many companies must also provide information on the environmental effects of their operations if they are engaged in activities that fall under environmental reporting requirements.

Larger companies also attach a *cash flow analysis report* to the annual report. As the name suggests, this analysis reports on, among other things, how operations were financed and which capital investments were made during the year. In short, the cash flow analysis shows how the company acquired and used capital during the year. The company's cash flow analysis is an important supplement to the balance sheet and income statement. See Chapter 8 for more discussion of financial analysis.

The auditor's report is the external auditor's analysis from the examination of the company's financial statements and the Board's and the CEO's management of the company. The auditor, who is appointed at the Annual General Meeting (AGM), primarily examines the company on behalf of its owners. In addition, however, the auditor checks that the company has reported and paid the correct amount of income taxes, employer contribution taxes, and VAT.

Major corporations and other publicly listed companies (e.g., companies listed on the Stockholm Stock Exchange) must appoint an *authorized public accountant* as the auditor. The auditor's reports for limited liability companies are publicly available. These reports include an opinion on whether the current accounts and the financial statements comply with generally accepted accounting principles and the law.

In addition, the auditor's report includes a statement on whether the Annual General Meeting (AGM) has any reason not to grant the Board of Directors and the CEO *discharge from liability* for the past year. If the auditor finds mistakes or other problems in the accounting records, the auditor notes them in the auditor's report so that they are brought to the attention of the AGM. Furthermore, the auditor has to send the auditor's report to the Tax Authority if serious errors are found in how the taxes and other charges were handled.

The limited liability company submits its annual report and its auditor's report to the shareholders at the AGM no later than six months after the financial year-end. The AGM approves the balance sheet and the income statement and decides on the *appropriation* of the company's *profit*. In addition, the AGM decides on the *discharge from liability* for the Board of Directors and the CEO, which means that they will not be held responsible if claims of mismanagement are brought against the company in the future.

Larger companies and other publicly listed companies must submit an *interim report* at least once a year. This report, which contains an income statement and key balance sheet items, describes the company's operations and financial results, including information on investing and financing and changes in liquidity since the previous year.

7.4 Preparation of the financial statements

Preparation of a company's financial statements involves a number of revisions and additions to the information in the company's bookkeeping records. Besides these revisions and additions, which we discuss below under the headings "Allocation of income and expenditures" and "Valuation of assets and liabilities (debts)", a number of control steps are required. Each account in the records must be reviewed and reconciled. This means checking the recorded amounts against invoices and other documents. Even for small companies, this is quite a time-consuming task. Generally speaking this work involves the following:

- allocation of income and expenditures to a particular period
- valuation of the assets and liabilities (debts)
- allocation to untaxed reserves.

Allocation of income and expenditures

During the year the company has recorded many items of income and expenditure in the books. These amounts can be verified by reference to invoices received and invoices sent. In preparing the financial statements, these amounts must be allocated to the correct period. *Allocation* means determining which income amounts agree with the "value of the delivered goods/services" during a specific period (i.e., revenues) and which expenditures agree with the "value of resources consumed" during a specific period (i.e., costs).

The allocations are of four types:

1. prepaid expenses
2. deferred income
3. accrued expenses
4. accrued income.

Prepaid expenses represent expenditures during a period but without an associated cost (i.e., payments for goods/services that have been paid in advance of their consumption or use). Typical examples are rents (generally paid in advance), and insurance fees (which may apply to both the current year and the next year). Prepaid expenses are assets on the balance sheet instead of costs on the income statement.

Deferred income represents income during a period without an associated revenue (i.e., payments recorded in advance for goods/services that have not yet been earned or delivered). A typical example is a customer prepayment. The company has recorded the receipt of cash as an asset, but is required at a later date to provide the service or deliver the product that has been paid for in advance. Deferred income is recorded as a liability on the balance sheet, not as revenue on the income statement.

Accrued expenses represent resources that have been consumed (i.e., costs) but have not yet been recorded because the company has not yet received an invoice for the goods/services purchased. Therefore, the company has accrued the cost, but has not (yet) recorded the associated expenditure. A typical example is a telephone invoice that is paid quarterly. For example, an invoice from the telephone company may overlap two accounting years; some of the amount on the first invoice received in the next year may relate to the current year. In addition, often many invoices have not been received at year-end although they are costs for the current year. The company either knows the exact amount or must estimate it in order to record these accrued costs. Accrued expenses are liabilities on the balance sheet and costs on the income statement.

Accrued income represents revenues earned that the company has not yet invoiced. Therefore, the company has not been paid. The simplest example is the situation when a company has performed a service or delivered a product but has not yet sent an invoice to the customer. The company usually knows the exact amount of the accrued revenues, which are assets on the balance sheet and revenues on the income statement.

In bookkeeping, one often comes across the term *interim claims* in reference to financial statement preparation and the allocation of prepaid expenses and accrued income to the correct accounting period. In the same way, one finds *interim liabilities (debts)* in reference to financial statement preparation and the allocation of deferred income and accrued expenses to the correct accounting period.

Figure 7.9 diagrams these four concepts.

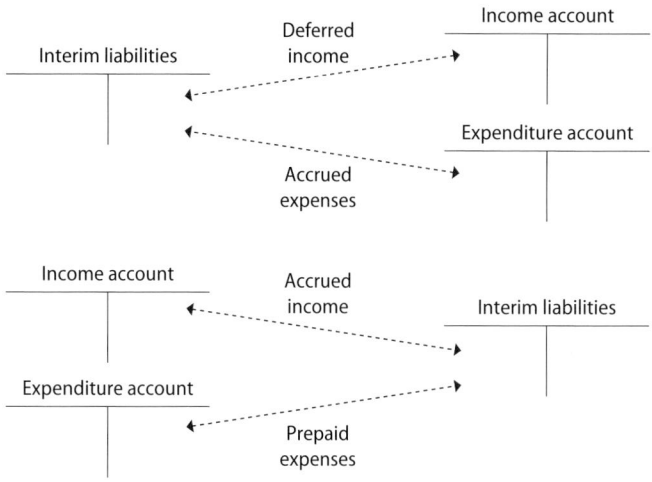

Figure 7.9 Coding of accrued and prepaid/deferred expenses and income.

Valuation of assets and liabilities (debts)

As we recall from Chapter 6, a company's profit can be determined in two ways – first, by subtracting costs from revenues; second, by calculating the difference between the change in assets (beginning of year to end of year) and the change in liabilities or debts (beginning of year to end of year). In other words, the year's profit appears on both the balance sheet and the income statement. Because the assets' recorded value in the bookkeeping (to be distinguished from their actual market value) is crucial in the determination of annual profit, it is necessary to value these assets as correctly as possible. *Asset valuation* is performed for both current assets and fixed assets, in part by different rules.

Sweden, like many countries, has two sets of laws for the valuation of assets; these laws have different purposes. First, *civil law* (e.g., the Annual Accounts Act), supported by the external auditor's examination, is intended to protect a company's stakeholders from over-valuation of the assets. Over-valuation of assets may mean, for example, that the company owners and financiers receive an inflated presentation of the company's financial position. Second, *taxation legislation* (the Tax Code) is intended to pro-

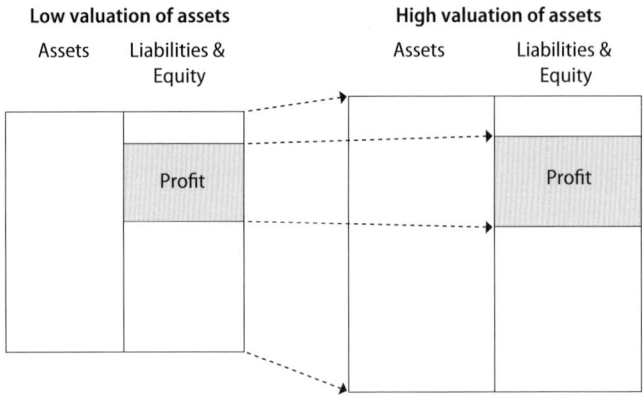

Figure 7.10 The influence on profit of asset valuations.

tect the government from under-reporting tax liabilities by companies. If a company under-values its assets, at least in the short term, it will report lower profit and thus pay less tax. See Figure 7.10. Compare with Figure 7.2 that illustrates how the valuation of the company's assets influences the calculation of profit (assuming debts and equity are unchanged).

We also observe that a company's liabilities need to be properly recorded. However, this task is generally easier than the valuation of assets because the amount of most liabilities can be traced to formal documents (invoices, loan contracts, etc.).

Fixed assets

A company's fixed (long-term) assets are those assets that are used over longer time periods (i.e., longer than a year). Valuing such assets involves the bookkeeping process of depreciation, which is a systematic method for allocating the cost of a fixed asset (e.g., a machine) to the years in which the asset is used in company operations. For example, consider a truck. The company purchases the truck at a considerable cost in Year 1 and expects to use it for a number of years (e.g., five years). The cost of the truck, therefore, is its depreciation in years 1 through 5. The Annual Accounts Act governs this calculation, which is called *depreciation according to plan*.

Different fixed asset categories are valued differently, predominantly because of the differences in their estimated economic lives. A building, for example, can be assumed to have a longer life than a truck. Buildings, therefore, are depreciated over longer time periods. In the following, we review the rules for *depreciation according to plan* for the four asset categories:

1. machinery and equipment
2. goodwill
3. real estate
4. land improvements.

The highest valuation allowed (by the Bookkeeping Act) for *machinery and equipment* is its cost of acquisition (i.e., the value it had when acquired). This determines the upper limit of its depreciable value. The assumption is that machinery and equipment continuously decrease in value according to a predetermined plan. If the useful (economic) life is estimated at ten years, the planned depreciation rate each year is 10 % of the cost. If the economic life is five years, the planned depreciation rate each year is 20 % of the cost. This is depreciation according to plan.

The Tax Code determines the lower limit of the depreciable value of machinery and equipment. This law allows companies to depreciate machinery and equipment at an accelerated annual rate. Thus, the company may report lower values for these assets in the financial statements than the amount of depreciation calculated according to plan, resulting in lower profit and, thus, lower tax (the tax rules are discussed in more detail below in connection with the possibilities for making appropriations). In some cases, equipment may be fully depreciated (at 100 %) in the first year, which means that the asset is treated as a cost in the year of its purchase. This is called accelerated depreciation for *consumables* (i.e., equipment that will be used up in less than 3 years, or equipment that is of low value – that is, half of the *price base amount* – in 2015; half of SEK 44,500, or less.

Goodwill is the difference between the amount of money a company paid for purchasing another company and the value at which the acquired company is recorded (known as the book value of the acquired company). In the purchaser's accounting records, the difference is recorded as Goodwill.

Goodwill is an asset on the company's balance sheet and is depreciated at the same rate as the asset's real value declines. When goodwill is depreciated (as a one-time event or over a period of years), it means the goodwill decreases on the balance sheet and it becomes a cost on the income statement.

Real estate is generally depreciated according to the Tax Authority's tax tables that prescribe different depreciation rates for different kinds of buildings. Thus, the annual rates may vary from 2 % to 5 %. Note that only the buildings are depreciated, not the land.

Land improvements, for example, parking lots and landscaping, are depreciated at 5 % a year. Note that undeveloped land is not depreciated.

The bookkeeping records typically use separate accounts for different asset categories. These assets are recorded at their cost of acquisition. Depreciation according to plan of these assets is recorded in separate accounts. For example, an account for accumulated depreciation of machinery (on the balance sheet) is used for the total of depreciation taken on the machinery. In a sense, one may think of accumulated depreciation as a negative asset, since the account is credited. Of course, there is also a cost account used for the annual depreciation amount (which appears on the income statement). Figure 7.11 illustrates how purchase and depreciation of a piece of machinery are recorded in the books.

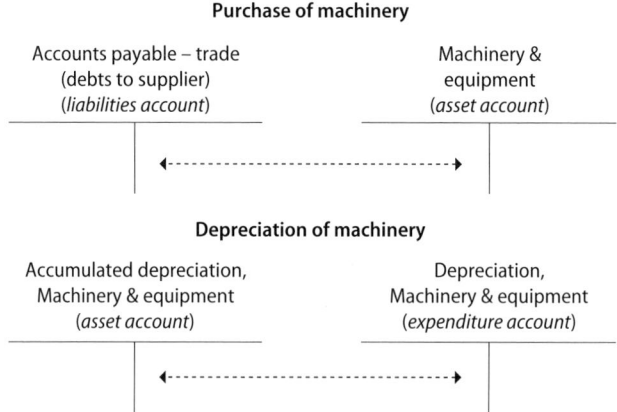

Figure 7.11 Example of accounting for the purchase and depreciation of machinery & equipment.

Current assets

The valuation of *current assets*, mainly *raw materials inventories, products in process,* and *finished goods inventories,* may pose some ambiguities. For other current assets, such as *cash on hand, bank balances,* and *accounts receivable,* precise information is usually available (e.g., from bank statements and invoices).

Inventories should be valued using the principle of *lower of cost or market.* This means that they should be valued at the lower of either their cost at acquisition, or their current market value. Cost at acquisition is normally the purchase price. Market value may be the estimated selling price at year-end, or an estimated replacement cost. In practice, inventories are usually expensed (depreciated) annually by 3 % for *obsolescence.* Obsolescence means that some of the inventory is unsellable or unusable because it is too old, out of date, or damaged.

In practice, the value of inventory is determined by a physical count of all raw materials, components, and products, which are listed according to types and then valued. Very often, this is an extensive process, especially at large companies, because there may be thousands of different materials and products in various stages of completion. The inventory count takes place at least once every year (in conjunction with the preparation of the financial statements), and more often if a company presents interim financial statements. A list of inventory items is prepared, and then the lower of cost or market rule is applied to items or groups of items.

Appropriations for untaxed reserves in the financial statements

Swedish tax law allows businesses, within certain limits, to reduce their reported profit through *appropriations.* These are reported as costs recognized in the income statement and create *untaxed reserves* on the balance sheet. See Figure 7.12.

Companies have the option to choose between two types of appropriations. The first type of appropriation is *depreciation as recorded in the books* (according to tax law) of certain assets. This typically results in larger depreciation than the *depreciation according to plan;* the difference between

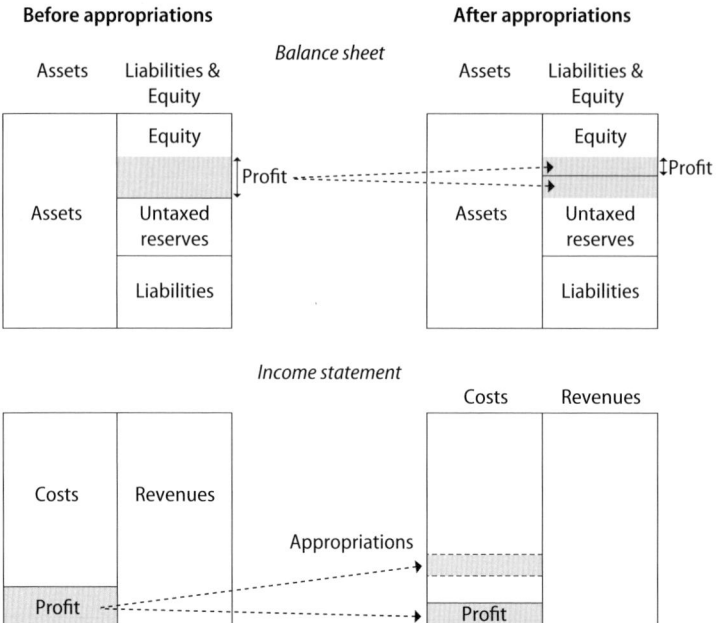

Figure 7.12 Appropriations and untaxed reserves.

these two depreciation calculations is an appropriation called *additional depreciation*. The second type of appropriation results from the allocation of profit to a *tax allocation reserve*, which can be made as a percentage of the profit before tax.

Additional depreciation

As discussed above, fixed assets are depreciated according to plan. Tax law, however, allows certain fixed assets to be depreciated more rapidly using *additional depreciation*. Figure 7.13 illustrates the link between depreciation according to plan and additional depreciation.

The assets eligible for additional depreciation are machinery, equipment, vehicles, ships, patents, rented apartments, and goodwill. As Figure 7.13 illustrates, the total depreciation determines the lowest allowed recorded value

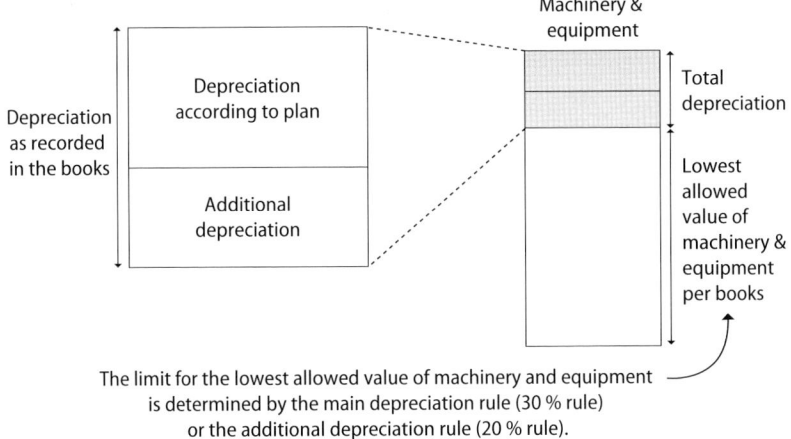

The limit for the lowest allowed value of machinery and equipment
is determined by the main depreciation rule (30 % rule)
or the additional depreciation rule (20 % rule).

Figure 7.13 Different depreciation concepts.

of these assets (machinery, equipment, etc.) in the bookkeeping records. This minimum value, in turn, depends on two rules: the *main depreciation rule (30 % rule)* and the *additional depreciation rule (20 % rule)*. Each rule determines an asset value, and the company may choose the lower of these values each year. However, the same rule must be used for all assets during the same year. Figure 7.14 shows how these values are calculated under the two rules.

The rationale behind the *additional depreciation rule* is that a company that follows the main depreciation rule can never reach a zero value for its depreciable assets. Even without additional investments, under the main bookkeeping rule, there is always a certain book value. However, if the company makes investments continuously in such assets, following the main bookkeeping rule will be more advantageous (i.e., will give the lowest value of the assets).

These two rules set the maximum limits of the depreciation (i.e., the lower limits on the asset's value). The amount of depreciation a company actually takes could be lower. In the bookkeeping, the *additional depreciation* amount appears in a cost account and the *accumulated additional depreciation* appears as a liability account (or negative asset).

Main depreciation rule (30 % rule)	Additional depreciation rule (20 % rule)
Book value of machinery & equipment beginning of the year	80 % of acquisition for machinery & equipment acquired in the year and remaining at year end
+ value of year's acquisitions of machinery & equipment	+ 60 % of acquisition value of machinery & equipment acquired in the previous year and remaining at year end
− value of year's sales of machinery & equipment	+ 40 % of acquisition value of machinery & equipment acquired 2 years ago and remaining at year end
= depreciation base	+ 20 % of acquisition value of machinery & equipment acquired 3 years ago and remaining at year end
The lowest allowed value of machinery & equipment, per books, is 70 % of depreciation base	= The lowest allowed accounting value of machinery & equipment, per books

Figure 7.14 Main depreciation rule (30 % rule) and additional depreciation rule (20 % rule).

The tax allocation reserve

In addition to *the additional depreciation*, a company is allowed to set aside a percentage of its profit before tax in the *tax allocation reserve*, which gives the company the opportunity to distribute its profits over several years. The allowed amount has varied somewhat. In 2015 the percentage was 25 %.

The transfer to the tax allocation reserve is the last step before the company's tax is calculated. All other costs, including additional depreciation and other adjustments, must be recorded before the transfer. The allocation is subtracted on the income statement in the same way a normal cost is subtracted. The allocated sum may remain in the tax allocation reserve for five years; after that, it is subject to taxation. After five years, the company simply adds the particular amount to its annual profit. However, a company may also take the allocated sum into profit at an earlier time, for example in a year when there is a loss. In this case, the company can use this untaxed reserve (it is an untaxed reserve because it consists of a share of untaxed profit) to cover a loss instead of using taxed capital, in the form of profit brought forward.

The bookkeeping for allocation to the tax allocation reserve is handled in much the same way as accelerated depreciation.

7.5 Taxed profit compared with accounting profit

Like private individuals, companies file an annual *tax return* with the Tax Authority. The *tax return* determines how much in taxes and other charges the company is required to pay. In addition to tax on profit (*corporate income tax*) the company also pays other taxes and fees (e.g., property tax and VAT). Examples of fees are *employers' social costs for labor* and *social security costs*.

When a company prepares its financial statements for the year, its corporate tax appears on the income statement as a cost. The item in the income statement, which is called "tax on profit for the year", is a calculated amount. In Sweden, in 2015, the corporate tax rate was 22 %.

Based on a company's tax return, the Tax Authority determines the exact amount of the tax owed. In addition to the corporation tax, the company must also pay (or recover) the difference between *Input VAT* and *Output VAT*, and pay any property taxes. Moreover, not all company costs are deductible in the tax return. Such costs, in whole or in part, may not be subtracted from revenues. Examples of such costs that cannot be claimed fully for tax purposes are some automobile leasing costs and some business entertainment costs (a maximum is allowed).

In addition to taxes on profit (corporate tax), on assets (property tax), and VAT, companies are required to pay fees of various kinds. These include, for example, *employers' social costs for labor* and *social security costs*, which are calculated as a percentage of salaries. In Sweden in 2015, these fees were approximately 45 % of most employees' salaries, which means that if a company pays an employee SEK 100 per hour, it effectively has to pay SEK 145 per hour. Like individuals, several of these charges, as well as some of the taxes (e.g., VAT), are paid periodically throughout the year.

Once the company has filed its tax return and the tax and fees have been calculated, the company records this amount as a *tax liability (tax debt)* in the current liabilities on the balance sheet. This liability is netted against the total of taxes and fees paid throughout the year. If an amount is due, this amount is then paid, and the tax account returns to a zero balance.

7.6 What do the financial statements tell us?

The financial statements present an overview of a company's performance and its consumption of resources in a certain time period, often a year (the income statement: revenues less costs) as well as a picture of its financial position at one point in time, often at year-end (the balance sheet: assets, liabilities, and equity). One way to analyze the financial statements is to use *business and financial ratios* in the *financial analysis*. Different business and financial ratios can, of course, be used to describe many company conditions as well as company finances.

In financial analysis, we often use three groups of ratios: *earnings ratios, liquidity ratios,* and *ratios of financial strength*. In using these ratios, it is important to remember that the ratios by themselves do not tell us much about a company. The ratios must be considered in relation to something, for example, the company's historic ratios, industry average ratios, competitors' ratios, etc. These key ratios are useful to us only when we use them in comparisons. Therefore, it is especially important in financial analysis to be aware of the company's type of businesses and its markets. Traditional business key ratios are, for example, not particularly useful for government agencies (which are not profit-seeking) or companies with monopolies (which have no competitors). Using these types of ratios to analyze a company is only relevant if the company has competitors.

Different companies present balance sheets and income statements that look rather different. In part, this is because they are in different industries. Table 7.1 shows the balance sheets of three companies that operate in three different industries: the mining company, LKAB; the truck manufacturer, Scania; and the IT consulting company, HiQ. Their balance sheet numbers (presented here as percentages) are from the years 2012 and 2013. By using percentages, we are better able to demonstrate the differences in the structure and composition of their balance sheets.

The first line item, *fixed assets* (non-current assets*)*, constitutes 57 % of LKAB's total assets and 58 % of Scania's total assets but only 40 % of HiQ's total assets. At first, these differences do not seem very large. An examination of these companies' comments (footnotes) to their annual reports is useful. We see that LKAB's fixed assets are mostly mines, roads, and installations

Table 7.1 Examples of balance sheets from three companies in three different industries: The numbers are percentages of the total capital. Source: LKAB, Scania, and HiQ.

	LKAB	Scania	HiQ
Fixed assets	57 %	58 %	40 %
Current assets	43 %	42 %	60 %
Total assets	100 %	100 %	100 %
Equity	45 %	31 %	72 %
Untaxed reserves	34 %		
Non-current liabilities	12 %	37 %	3 %
Current liabilities	9 %	32 %	25 %
Total equity and liabilities	100 %	100%	100%

needed to access ore in the mountains. Scania's fixed assets consist primarily of different interest-bearing assets (also called financial assets). These are predominantly customer leases of vehicles manufactured by Scania. In addition, Scania's fixed assets also include factories, machinery, and equipment.

For a consulting company, HiQ has a relatively high proportion of fixed assets. However, almost 90 % of these assets consist of intangible assets. A consulting company clearly does not require the extensive facilities (e.g., factories and heavy machinery) that a mining company or a vehicle manufacturer needs. HiQ's intangible assets are mostly *goodwill* – the difference between the price HiQ has paid for acquired companies and those companies' book value. Thus, HiQ has partly grown by acquiring smaller consulting companies, which affects the intangible asset goodwill.

The next line item is *current assets*. LKAB's current assets are inventories and a significant number of short-term investments (investments expected to be held for a year or less). Scania also has a number of short-term investments. About half of HiQ's current assets are accounts receivable (claims to payment for consulting contracts completed and invoiced, but not yet paid).

The third line item, *equity*, varies in size among the three companies. For example, Scania's equity is the lowest (31 %), and HiQ's is the highest (72 %). This means that Scania is 69 % financed by liabilities, and HiQ is

28 % financed by liabilities. One explanation for these differences relates to the willingness of banks and other financiers to lend money to companies, as well as to companies' attitudes toward borrowing. For example, the banks are relatively willing to lend money to LKAB because it is a company owned by the government. Therefore, lenders are at less risk. Furthermore, as noted above, LKAB's fixed assets are tangible and therefore relatively easy to value. Finally, LKAB's business is rather secure given the existence of the iron ore and the market demand for iron. By contrast, HiQ's is not government-owned and has few tangible assets. Its main asset, the knowledge and experience of its employees, is not even included in the balance sheet.

Earning capacity (return on capital and profit margin)

Earning capacity (or profitability) is a measure of a company's economic performance in terms of profit and loss in relation to some other measure. If a company's earnings (i.e., its profit) are compared to its capital (e.g., its equity or total assets), we can calculate various ratios in order to find out how *profitable* the company is.

There are many such ratios used for different purposes. Different authors and financial analysts use slightly different definitions in the numerator and denominator of the equations. It is important to carefully examine exactly how a specific ratio is calculated. In the following, we briefly discuss four of the most commonly used earnings ratios:

1. return on equity, ROE
2. return on total capital (or total assets), ROT
3. return on capital employed, ROCE
4. profit margin.

Return on equity expresses the return on the shareholders' capital in a company. It measures the company's ability to generate profit in relation to the portion of the company owned by the shareholders. This calculation is made using the income statement profit after financial revenues and costs over adjusted equity.

$$\text{ROE} = \frac{\text{Profit after financial revenues and costs}}{\text{Adjusted equity}}$$

Adjusted equity is equity plus the part of the untaxed reserves that are assumed to belong to the owners (i.e., equity plus untaxed reserves · (1 – tax rate)). This number is used as the capital measurement in the denominator. Sometimes an *average adjusted equity* is used (i.e., the average of this year's balance sheet and last year's balance sheet). In addition, depending on the purpose, the profit in the numerator can be calculated before, or after, tax.

Return on total capital (or total assets) expresses the company's return on total capital invested (i.e., the profitability on all capital, not only equity). For this calculation, financial costs (e.g., interest costs) are not considered because these costs can be seen as payments for financing the company. Here also, an *average adjusted total capital* is often used (i.e., the average of this year's balance sheet and last year's balance sheet).

$$\text{ROT} = \frac{\text{Profit after financial revenues}}{\text{Total capital}}$$

Return on capital employed (ROCE) is calculated in a similar way. However, capital that a company does not pay interest on is not considered in this ratio (e.g., its accounts payable – trade, which are the amounts owed to suppliers). There is no generally accepted definition of this measure. Different definitions are used in different contexts.

Finally, *profit margin* is also a common earning capacity ratio, even if it formally has nothing to do with the company's profitability. The profit margin relates the annual profit to the turnover (i.e., total net sales, which are sales excluding VAT). The profit margin is usually calculated by adding interest revenue and costs to the operating profit (i.e., financial items are considered generated from the operations as well). Through the profit margin, the size of the profit is related to the total size of the business.

$$\textbf{Profit margin} = \frac{\text{Profit after financial revenues and costs}}{\text{Turnover}}$$

The concept of profit margin is sometimes also used to analyze the profitability of specific products or services. See Chapter 4. On these occasions, the profit margin is calculated in a similar way as above (i.e., the difference between the product price and the total costs of a product, divided by the product price).

Liquidity ratios

Liquidity refers to the relationship between a company's *current assets* (or portion thereof) and its *current liabilities* (or portion thereof), or its *turnover*. Liquidity ratios give us an idea of the company's ability to pay its invoices in the short term. The term solvency is also used to describe a company's liquidity.

Two commonly used liquidity ratios are the following:

1. Acid test ratio
2. Current ratio.

The *acid test ratio* expresses the relationship between a company's most liquid assets and its current liabilities. The ratio should be at least 1.0, which means there are sufficient current assets to pay the current liabilities, if circumstances so require.

$$\textbf{Acid test ratio} = \frac{\text{Current assets – Inventories – Prepaid expenses}}{\text{Current liabilities}}$$

The second liquid ratio – *the current ratio* – sets all the current assets (including the inventories and prepaid expenses) in relation to the current liabilities. As a rule of thumb for a manufacturing company, this ratio should be at least 2.0.

$$\textbf{Current ratio} = \frac{\text{Current assets}}{\text{Current liabilities}}$$

Another ratio sometimes used is *cash and bank assets* over *total turnover (sales)*. In general, a company's need for liquid assets depends on its size. Thus, a company with higher turnover generally requires more current assets than a company with lower turnover.

Financial strength – Equity ratios

The Equity ratio (also referred to as financial strength) is a calculation in which a company's equity is compared to its total liabilities (debts). This calculation is an indication of how stable the company is should it incur future losses because losses are charged against equity. A company with a high equity ratio, for example, can more easily tolerate higher production costs or new market expansion costs than a company with a lower equity ratio. In fact, a company with a low equity ratio may be unable to take such actions.

In Sweden, three commonly used equity ratios are the following:

1. Equity ratio 1
2. Equity ratio 2
3. Debt/equity ratio.

Equity ratio 1 sets the adjusted equity in relation to the company's total capital. Thus, the ratio illustrates the owners' share of the company.

$$\textbf{Equity ratio 1} = \frac{\text{Adjusted equity}}{\text{Total capital}}$$

As above, adjusted equity is equity plus the part of the untaxed reserves that are assumed to belong to the owners (i.e., equity plus untaxed reserves · · (1.0 – tax rate)).

Equity ratio 2 is a variation of the same ratio. Here, *risk-bearing equity* is used in the numerator. Risk-bearing equity includes 100 % of the untaxed reserves. The reason is that the company must first meet possible losses by using such reserves before they can charge them against equity.

$$\text{Equity ratio 2} = \frac{\text{Risk-bearing equity}}{\text{Total capital}}$$

Debt/equity ratio sets liabilities or debts (including the deferred tax debts in the untaxed reserves) in relation to the adjusted equity (i.e., equity plus untaxed reserves · (1.0 – tax rate)).

$$\text{Debt/equity ratio (D/E)} = \frac{\text{Liabilities} + \text{Untaxed reserves} \cdot \text{tax rate}}{\text{Adjusted Equity}}$$

Other ratios

Various financial conditions can be investigated based on information in a company's financial statements. For example, it is often of interest to examine how much of the company's capital is invested in various current assets. The amount of a company's inventory and its accounts receivable may give us an indication of how efficiently the company conducts its operations. It may be of interest to compare the total of accounts receivable to the total of the company's annual turnover. For instance, if accounts receivable are SEK 5 million, and annual turnover is SEK 60 million, it means, on average, it takes the company about one month to collect payments from its customers. Depending on the purpose, many other ratios can be calculated. As mentioned before there are no strict and commonly used definitions of financial ratios.

7.7 Consolidated financial statements

A *group* of companies consists of a parent company and one or more subsidiaries, wholly or partially owned. The *parent company* owns, or controls, more than 50 % of the voting rights of its *subsidiaries*. The subsidiary may in turn have control over other companies, which the parent company then indirectly controls. Figure 7.15 is an example of such parent-subsidiary relationships. As the figure shows, the parent company directly owns three subsidiaries, and one subsidiary owns a subsidiary. Thus, there is a group within the group.

Most large industrial companies, including most of Sweden's publicly listed companies, are parent companies in parent-subsidiary groups. Since the mid-1900s, there have been many business mergers and acquisitions in Sweden and around the world. Economies of scale and synergies in operations, as well as internationalization and globalization of markets, are common explanations for this surge. As a result, the Swedish business community has become increasingly concentrated, especially in mature industries, as companies have bought their competitors and other companies have merged.

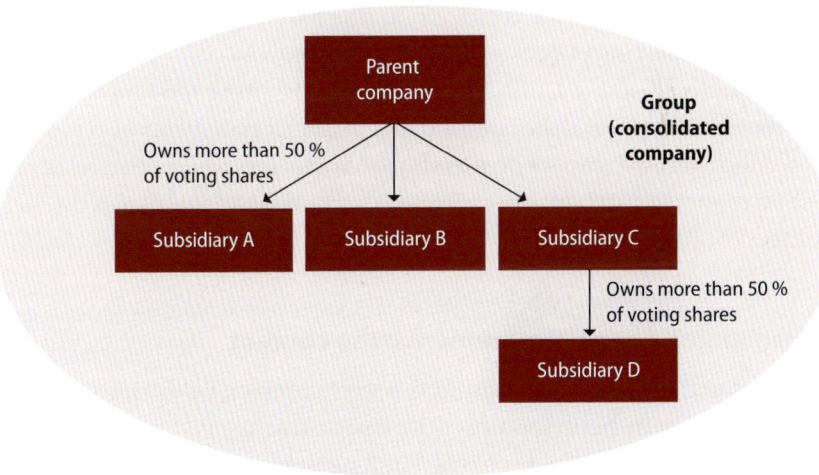

Figure 7.15 A group (consolidated company).

The group is a legal concept. Extensive legislation exists related to the creation of groups, their structure, and their annual financial reports. There is thus an important difference between subsidiaries in a group and the divisions in a single company. A division is an organizational concept. No legislation or mandatory rules govern its creation or structure.

Why do we have consolidations accounting?

After the notorious *Kreuger crash* in Sweden in the early 1930s, it was apparent that external stakeholders needed more reliable financial information on large business groups. Ivar Kreuger, a Swedish industrialist, had created a parent company, Kreuger & Toll, with over 600 subsidiaries. Behind this activity, among other things, was Kreuger's intent to hide losses and fabricate profits by submitting false invoices between companies under his control.

Through *transfer pricing* mechanisms and *group contributions*, a parent company can, in principle, decide where profits or losses should be recognized in the group. Consequently, the purpose of consolidated financial statements is to provide the stakeholders with a clear and comprehensive picture of the entire group. The transactions (e.g., purchases and sales) between members of the group are eliminated in the consolidated financial statements. In this way, the entire group (the parent and its subsidiaries) is treated (for reporting purposes) as one large company.

Legislation also prevents groups from distributing too large sums to shareholders. Therefore, the consolidated financial statements also provide protection against excessive dividends to shareholders who, with a relatively modest capital investment, can control a large number of companies through indirect ownership.

Consolidated financial statements: the content

Consolidated financial accounting is a very complex field. Here we only briefly describe its main features. The basic principle is that the consolidated annual report, as far as possible, imitates how the financial statements would have appeared if the consolidated group had actually been a single limited liability company. As mentioned above, this means, among other things, that all purchases and sales between the parent and its subsidiary companies, and between subsidiaries, are eliminated in the consolidated financial statements.

The group's annual report should, therefore, contain the same elements that a normal company's annual report does:

- consolidated balance sheet
- consolidated income statement
- administration report
- cash flow analysis (for larger groups).

In addition, each company in the group (parent and subsidiaries) prepares its individual annual report and submits a tax return on its profit. The parent company's individual financial statements are also included in the group's annual report.

Profits may be transferred between companies in the group if the parent company owns at least 90 % of the voting rights of the subsidiaries. This is called the *group contribution*. The rules for group contributions are designed so that the group's total tax liability is not larger than it would have been if the group had been a single company. We illustrate this idea in the following example.

EXAMPLE

A group consists of a parent company and a subsidiary. The parent company shows a profit of SEK 500,000 before tax for the year, while the subsidiary shows a loss of SEK 200,000 for the same year. If the group were only a single company, the profit would have been SEK 300,000. Without the rules for group contributions, the parent company would have to pay tax on its profit of SEK 500,000 while the subsidiary would owe no taxes. Thus, in that scenario, the total tax liability would be higher than it would have been without the group contribution. In practice, then, the parent company can make a group contribution of SEK 200,000 to the subsidiary to cover its loss. As a result, the parent company shows a profit of SEK 300,000, and pays tax on that amount.

The basic principles behind consolidated financial statements suggest their preparation is simple and straightforward. In practice, this is not the case. There are special valuation rules for the consolidated financial statements. The most important rule is that the parent company's valuation principles are used throughout the group. Thus, the subsidiaries' assets and liabilities are recalculated in accordance with how the parent company values its assets and liabilities. The Annual Accounts Act specifies how subsidiaries' assets and liabilities should be valued, how subsidiaries' revenues and costs should be recognized, and how acquisitions of subsidiaries should be accounted for.

CHAPTER 8

Corporate finance

In most industrial operations, there is a time lag between payments (cash out) and receipts (cash in). As an example, a company pays for raw materials; then these materials are put into production; finally, the finished products are sold. After the sale, the company receives cash. Another example is an investment in a new machine that the company plans to use for a number of years. The company pays cash for the machine at the time of purchase; then the company collects cash receipts over the years as the machine is used to manufacture products. Of course, there are exceptions, but the point is that generally companies have money (financial capital) tied up in materials, machinery, buildings and more. To finance its operations, the company must somehow acquire this financial capital.

Corporate finance can be discussed from two perspectives:

1. from the view of the company that requires capital to conduct its operations; or
2. from the view of the external actors who provide the company with financial capital by share purchases, loans, and other forms of financing.

In this chapter we focus on the first perspective. The second perspective, which has to do with venture (risk) capital and loans, is the subject of financial economics and is discussed extensively in specialized literature.

The way in which a company is financed is revealed on the liability/equity side (passive side) of the balance sheet. Chapter 7 reviewed the company's financial statements. In this chapter we analyze how to calculate the capital requirements of a company and its capital structure, including a more detailed breakdown of its liabilities. In our discussion of how a company can raise financial capital for operations, we illustrate how financial analysis is used to calculate and analyze a company's financing requirements. Finally, we discuss the risks associated with financing a company.

8.1 Capital requirements: Working capital

Most businesses need various types of financial capital to carry out their operations. Capital is needed to acquire equipment, hire staff, rent premises, purchase raw materials and components, engage consultants, and to create a financial safety margin to withstand unexpected events. Thus, cash budgeting, that is, liquidity planning, is crucial.

How much capital is required and how that capital is used vary, depending on the type of business. Some companies tie up extensive capital in fixed assets, such as plants, manufacturing equipment, and real estate. Examples of such companies are found in industries like mining, forestry, pulp and paper, chemicals, telecommunications, and power generation. Other companies, such as retailers and wholesalers, tie up most of their assets in inventories required to provide customers with the necessary goods and services. Consulting firms, on the other hand, seldom have inventories of this kind. Instead a significant part of their capital is commonly tied up in client projects that have been carried out but have not yet been paid for (in other words, they report accounts receivable).

The capital required to run a business thus arises because companies often need to make large payments in order to produce and sell goods and services long *before* the receipts (i.e., payments from customers) are received. Three common types of such capital are:

1. *Fixed asset capital:*
   to invest in assets for permanent use, for example, machinery, equipment, and real estate.
2. *Working capital:*
   to finance ongoing operations, that is, purchase of various goods and services used in the production process, such as raw materials, components, and consumables, and the cost for inventory of finished goods and for products that have been delivered but not yet paid for.

3. *Safety capital:*
   to create a buffer to cope with disruptions and unforeseen problems, usually a liquidity reserve in the form of cash or bank deposits.

Fixed asset capital requirement

The need for fixed asset capital is, at least in theory, relatively easy to determine. Basically, this capital is equal to the original investment in the classic investment calculation.

In practice, however, there are several examples of large construction projects that become significantly more expensive than originally calculated. In addition, major investment projects are often commercially risky and may take a long time to implement, which means that it often takes a long time before the company begins to receive receipts (payments from clients) on an investment. It is, among other things, this risk that companies try to take into account when they establish their cost of capital interest rate to be used in their investment appraisals.

One way to reduce or eliminate the need for fixed asset capital is to rent, or lease, premises and equipment instead of buying and owning them, thus avoiding the need to finance the large initial investment. At the same time, such solutions might be difficult to identify and they often mean higher overall costs in the long term. In addition, there might be several reasons why a company actually wants to own, and thereby completely control, their facilities.

Working capital requirement

It is, however, not enough just to finance the fixed assets. Companies also tie up capital in their current assets, that is, in the ongoing operations. The need for this capital arises because companies often have to make significant cash payments long before they get paid for goods and services delivered. It takes capital to acquire raw materials and components, pay wages, rents, etc., and these payments must be made even if the company has not yet received any receipts from sales of its products. The company must meet these *working capital* requirements because of this time-lag between payments and receipts. A lack of working capital, and the subsequent inability to pay the bills, is one of the most common reasons that companies file for bankruptcy.

An established company, with ongoing activities, can usually finance at least part of the operations with money from its sales. However, additional financing of parts of the working capital is often needed since large amounts of capital can be tied up in the ongoing operations. In companies producing goods, working capital is needed when raw materials and components are purchased and stored in *inventories*, during manufacturing of the products (*work in progress*, WIP), in inventories between production steps, in *inventories for finished goods* waiting to be sold and delivered to customers, and sometimes in road transport of goods to customers. In addition, it may take time before the company is paid. If sales are not made in cash, the buyer usually has a certain credit time, for example a month, which means that working capital is also tied up in the company's *accounts receivable* (i.e., invoiced amounts that have not yet been paid). Similarly, *accounts payable* (i.e., the company's unpaid invoices) reduce working capital requirements. This is also true for advance payments from customers.

At the end of the fiscal year, during preparation of the annual financial statement, it is necessary to obtain an inventory of how much capital is tied up in current assets, that is, inventories, work in progress, accounts receivable, etc. and the current liabilities, that is, accounts payable and advance payments from customers.

Capital tied up and capital rationalization

The level of *capital tied up* might thus be significant in many operations. The capital tied up needs to be financed, which in turn often leads to interest expenditures for loans and other debts. Moreover, the capital tied up in production processes, inventories, etc. poses business risks, since material, components, and products may become obsolete. Many companies, therefore, actively work to minimize the capital tied up in operations without decreasing production efficiency or the ability to deliver to customers. This is called *capital rationalization* and can be accomplished in several ways, for example:

- to keep inventories as small as possible
- to have as short lead times in production as possible

- to receive frequent just-in-time deliveries from suppliers, instead of keeping large inventories of raw materials and components
- to produce to customer order rather than selling from an inventory of finished goods.

Another way to exercise skill in business negotiations is to contract for long credit periods with suppliers and short credit periods with customers. In addition, the billing and payment systems should be designed so that the suppliers' credit terms are maximized while receipts from customers are received as early as possible.

As a consumer, you might observe how different companies have adapted their operations to reduce capital tied up. Retail stores such as Lidl and IKEA, for example, keep large parts of their inventories among customers, inside the stores. Many stores for consumer electronics regularly run large advertising campaigns for certain products in order to sell them *before* the invoices from the suppliers of the products must be paid. Another example is designer furniture stores. If you buy an expensive sofa or a dining room table, it may take several months from order and payment until the furniture is delivered. Such furniture is usually manufactured to order.

In an industrial company, a large part of the engineering work in production, logistics, and sourcing is focused on material flows and capital tied up. For a long time, the aim has been to reduce all types of inventory in production as far as possible, for example by not producing anything until the products are ordered and allowing component deliveries to take place just in time, that is, exactly when they are needed, and by obtaining a rapid flow through to production. Many large corporations, for example the major automotive manufacturers, have therefore transferred the responsibility for storing inventory of many components to their suppliers.

Consequently, it is important not to tie up more capital than necessary. The efficiency of *capital utilization* is typically reviewed and assessed regularly by various key performance indicators, such as inventory as a percentage of turnover, inventory turnover rate (how many times the inventory is sold over a period of time), average shelf life, average credit times, amount of work in progress, and turnover of accounts receivable.

Calculating working capital requirements

Working capital is thus, by definition, the capital tied up in operations that cannot be financed using the current (short-term) liabilities (primarily accounts payable). Instead, this capital requirement needs to be covered by long-term debt and/or company equity. Working capital is thus defined as follows:

Working capital = Current assets – Current liabilities

The working capital requirement, of course, differs for different types of operations. Figure 8.1 describes where working capital typically is tied up in different businesses and how the working capital requirement can be calculated. Retail companies mainly have working capital tied up in inventory and accounts receivable, while manufacturing companies in addition have much capital tied up in their production processes. Most service companies are not as capital intensive, even though there are exceptions. In these companies, most capital is tied up in work already performed in various client projects that have not yet been paid for.

We will now briefly describe how to calculate the working capital requirement in a manufacturing company. The model applies to all types of businesses, but a goods-producing company provides the most comprehensive calculation.

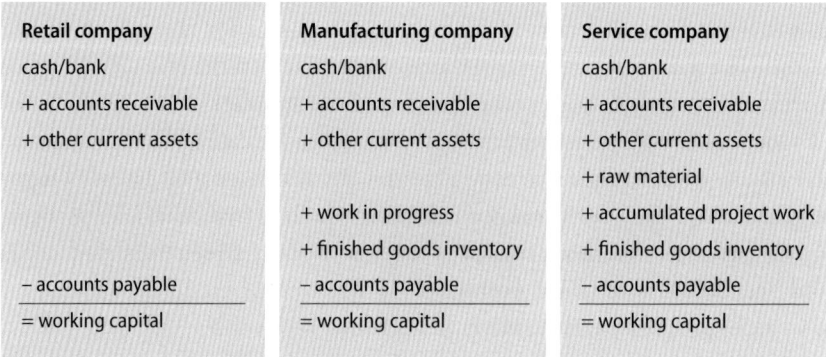

Retail company	Manufacturing company	Service company
cash/bank	cash/bank	cash/bank
+ accounts receivable	+ accounts receivable	+ accounts receivable
+ other current assets	+ other current assets	+ other current assets
		+ raw material
	+ work in progress	+ accumulated project work
	+ finished goods inventory	+ finished goods inventory
– accounts payable	– accounts payable	– accounts payable
= working capital	= working capital	= working capital

Figure 8.1 Difference of capital tied up in different types of operations.

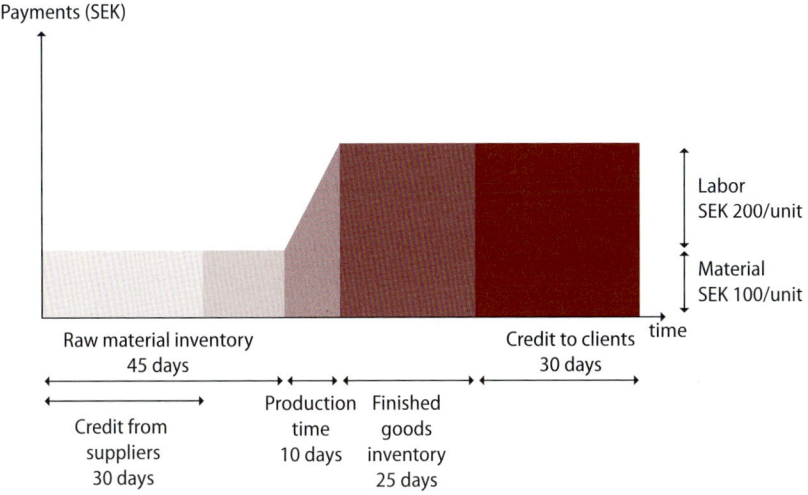

Figure 8.2 Example – Working capital tied up in a manufacturing company.

Figure 8.2 provides an example of how working capital can be tied up in a manufacturing company. In this company, capital is, on average, tied up in the following way: raw material is stored for 45 days, supplier credit time is 30 days, production time is 10 days, finished goods are stored for 25 days, and the credit period offered to customers is 30 days. Thus, the company does not receive receipts for the finished goods until 80 days after the raw materials need to be paid for.

During the production process, manufacturing costs are allocated to the products, making them increase in value so that the finished goods inventory is valued much higher than the raw materials inventory. Having finished products in stock is therefore very expensive. In the example, the material in the raw materials inventory is worth SEK 100 per product unit while the finished products are worth SEK 300 per unit. If the company produces 300 products per day, we can calculate how much working capital is tied up in these operations. The working capital requirement can be calculated in two ways: (1) the time method and (2) the balance method.

Time method

Capital tied up = Number of days · Cost per unit · Production volume per day

Days material = (45–30) + 10 + 25 +30 = 80 days

Days work in progress = $\frac{10}{2}$ + 25 + 30 = 60 days

Capital tied up material = 80 · 100 · 300 = SEK 2,400,000

Capital tied up work in progress = 60 · 200 · 300 = SEK 3,600,000

Capital tied up = SEK 6,000,000

Balance method

Capital tied up = Number of days . Cost per unit . Production volume per day

Raw material inventory = 45 · 100 · 300 = SEK 1,350,000

Credit from suppliers = –30 · 100 · 300 = SEK –900,000

Work in progress = 10 · 100 · 300 + $\left(\frac{10}{2}\right)$. 200 · 300 = SEK 600,000

Finished goods inventory = 25 · (100 + 200) · 300 = SEK 2,250,000

Credit to clients = 30 · (100 + 200) · 300 = SEK 2,700,000

Capital tied up = SEK 6,000,000

Fundamentally, the total area under the curve in the figure is calculated and adjusted for the supplier credit time (which reduces the capital tied up) and then multiplied by the production volume. A basic assumption in the model is that the value of the raw material inventory, the work in progress, and the products in the finished goods inventory are actually matched by payments of equal amounts.

The company in the example thus ties up SEK 6,000,000 of working capital in its operations. This means that if it is a newly established company it needs to invest SEK 6,000,000 in payments for raw materials, wages, etc. before it receives any revenues whatsoever. Alternatively, we might say that the company continuously needs to achieve a return on its operations that is large enough to finance working capital of at least SEK 6,000,000 in loans and other capital contributions.

Useful formulas

$$\text{Capital turnover rate } [\tfrac{\textbf{times}}{\textbf{year}}] = \frac{\text{Sales [per year]}}{\text{Total capital (at year's end [SEK])}}$$

$$\text{Inventory turnover rate } [\tfrac{\textbf{times}}{\textbf{year}}] = \frac{\text{Turnover period [365 days]}}{\text{Average time in inventory [days]}}$$

$$\text{Average credit time to clients [days]} = \frac{\text{Average credit to clients [SEK]} \cdot 365 \text{ [days]}}{\text{Sales [per year]}} =$$

$$\frac{\text{Average accounts receivable – trade [SEK]} \cdot 365 \text{ [days]}}{\text{Sales [per year]}}$$

Average credit time from suppliers [days] =

$$\frac{\text{Average credit from suppliers [SEK]} \cdot 365}{\text{Purchase costs [per year]}} =$$

$$\frac{\text{Average accounts payable – trade [SEK]} \cdot 365}{\text{Purchase costs [per year]}}$$

8.2 Capital structure

A company's capital structure is of great importance in decisions about financing. The three main financial sources for a company, in terms of time to maturity, are the following:

1. equity (including untaxed reserves)
2. interest-bearing liabilities (debts)
3. operational liabilities (non-interest-bearing debts).

A company's equity, including its untaxed reserves, and some interest-bearing liabilities are at risk if a company defaults in any way. In Sweden, the term *risk capital* is sometimes used to stress the risky nature of ventures in business. Companies that specialize in providing equity financing to high-risk business are also called venture capital firms for the same reason.

Equity

Equity, which is the owners' stake in the company, is the basis of the company's capital structure. Equity includes share capital, the retained earnings (i.e., earnings from previous years), and, in a financial context, a proportion (1.0 – tax rate) of some or all of the untaxed reserves. Previously (in Chapter 7) we used the term *adjusted equity* that refers to the *risk-bearing equity* and the untaxed reserves. Equity is the most long-lasting financing source because it exists as long as the company exists. The company owners, who supply this specific capital, take the greatest risk in the event of company default. A default may result in a restructuring of the company or in its bankruptcy. In either case, some or all of the equity is lost.

As we recall from Chapter 7, which dealt with the company's annual report, in the event of an annual loss, the loss is first offset by the company's untaxed reserves (i.e., part of equity). Because of the risk of losing their invested capital, owners also require the highest return on their investment (in particular, because of the *double taxation* of company profit and of shareholder dividends). From the company's perspective, when it calculates its *cost of capital*, equity has the highest cost of all the alternatives. See Figure 8.3.

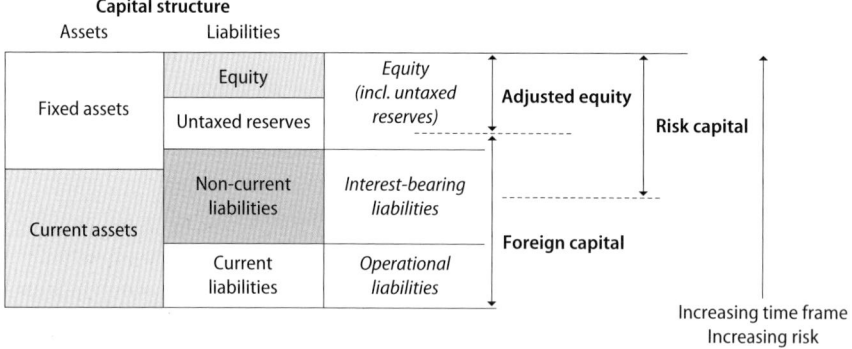

Figure 8.3 The company's capital structure.

Shareholders' equity (owners' equity)

It is essential to understand the capital structure of a company if we are to understand how power is distributed. In a limited liability company, ownership is divided equally by a system of *shares* that owners can purchase (and sell).

Shares of publicly listed companies can be bought and sold freely (not counting transfer costs) in the financial markets. Shareholders in unlisted companies can place limitations on how shares can be disposed of. All limited liability companies must maintain a *shareholders' register* of the names of shareholders and the number of shares they own. Only shareholders listed in the register are entitled to the stated shareholder rights (e.g., voting rights and dividend payments).

A company's share capital is the sum of the *quota value* (previously, *nominal value*) of all issued shares. Historically, the nominal value was printed on shares; this is no longer the case in Sweden where share certificates are no longer issued. All company shares in a given class have the same quota value and the same rights. However, a company may issue different classes of shares, distinguished by different *voting rights*, various *dividend preferences*, or certain share *transfer restrictions*.

- In Sweden, shares with different voting rights are usually referred to as *Class A shares* and *Class B shares*. Today, the difference in voting rights between the two classes cannot be more than 10 times; larger differences were permitted historically.
- Shares with certain preferences as far as dividends and distribution of assets are called *preferred shares*; other shares are called *common shares*.
- Previously there were also restrictions on share purchases, such as that only Swedish citizens could buy shares in Swedish companies. These shares were called *restricted shares* as distinguished from *unrestricted shares*. This restriction is very unusual today in Sweden.

This division of shares with different voting rights was once common in Sweden given its industrial and ownership structure with large companies and major investors exerting great influence. By issuing different classes of shares, large companies could raise capital while the major owners still maintained control. Furthermore, shareholders with preferred shares, with their special rights, could reduce their risk.

Other types of financial capital

A company's basic capital consists of its equity, all its untaxed reserves, and its *subordinated* liabilities. In the event of company bankruptcy, primarily the financiers who made interest-bearing loans to the company under subordinate terms are at risk.

Unlike ordinary interest-bearing debts, subordinated liabilities are ranked behind other liabilities as far as payment in the event of the company's bankruptcy. A typical type of subordinated liability is a *convertible loan*, which is a liability that can be exchanged for company shares according to an advance contractual agreement.

Although the basic capital involves more than just the equity (the share capital), the real power in a company belongs to the owners of the equity: the shareholders. They control the company – its formation and its operations. Nevertheless, the company's other financiers exert a strong influence on the company.

Because the lenders with subordinated liabilities have a lower priority than other lenders, they are at greater risk. Therefore, they naturally require a higher rate or return (i.e., interest rate) than the lenders with unsubordinated liabilities.

Interest-bearing liabilities

Interest-bearing liabilities are loans in which the cost of capital takes the form of an interest rate. A company borrows a certain amount of money (e.g., from a bank), and makes interest payments on the loan. Typically, the bank requires some form of security for a loan. For example, with mortgage loans, a real estate property can be the security. This means, in the event of a company default on the loan, the bank has the legal right to the property. Given this arrangement, the bank may set an interest rate that is less than the owners' required return on their investment.

Operational liabilities

Operational liabilities (current liabilities on the balance sheet) consist mainly of liabilities the company has to its suppliers. Because such liabilities are usually interest-free, they are the cheapest form of company financing. However, the maturity dates of operational liabilities are usually quite short (e.g., a few weeks or a month).

Matching of assets and liabilities

A basic principle in all business financing is matching the *maturity date* of the financing source to the financed asset's life. One usually refers to matching the maturity date of a liability to the economic life of an asset. The maturity date is the date when payment of the loan principal is required.

For reasons related to liquidity, the total of a company's most liquid assets – its current assets – should be greater than the total of its operational (current) liabilities. That is, the company should have a liquidity position in which its *acid test ratio* is greater than 1. Otherwise, the company will likely experience short-term payment problems. Preferably, current assets should be financed by medium-term current liabilities. Other assets (i.e., the fixed assets) should be financed with long-term liabilities, untaxed reserves, and equity. The ideal balance between financing sources and financed assets depends on the nature of a company's operations. If the business is capital intensive (e.g., the pulp and paper industry), asset financing must be long term. If the company produces and sells consumer electronics with frequent launches of new products in the market, the financing considerations must allow for frequent replacements of production equipment as a consequence of the product changes.

8.3 Financing the operations

All the company's purchases are paid for, more or less directly, with liquid funds. When a company buys a new machine, cash (a liquid asset) becomes a fixed asset (the machine). Maintaining a reasonable level of liquid assets is, therefore, a fundamental principle of financing. A second fundamental principle is designing an appropriate structure of liabilities for the company's operations while still maintaining a reasonable level of equity. Here, we review the most common sources of financial capital (both internal share capital and external capital).

Generating capital from operations

The annual profit is a part of the company's equity. This means that a company generates capital for its operations through its earnings. Self-generated capital is one of a company's most important sources of capital. Terms often used to describe this method of capital creation are *organic growth* and *self-financing*. In a situation where the company grows without a change in its equity and debt (a constant equity ratio), the term used is *growth under financial balance*. If a company wishes to maintain growth under financial balance, its total assets cannot grow more rapidly than its equity. This means that *the return on equity* sets the limit for how fast the company can grow. If the company intends to grow faster, without decreasing its equity ratio, the owners must make some form of financial contribution.

It is important to observe that a very profitable company does not necessarily generate surplus cash. Later in this chapter, when we discuss financial analysis, it will be clearer how the company's profit (or loss) and the change in its operational liabilities are linked. At this point, we will only remark that changes in liquid assets, of course, depend on how the company used its financial capital during the year, and on how much capital came into the company from various sources (the profit is only one source). A company that is very profitable in one year, and at the same time makes significant investments in fixed assets, may actually decrease its liquid assets.

Issuing shares

If a company is unable to self-generate enough financial capital, for example, to increase the scale of operations or to initiate a start-up phase, it can seek financial contributions from existing owners and from potential new owners in the form of *share issues*. The most common way to increase the share capital is to launch a new share issue. See Figure 8.4.

A share issue means that a company issues a number of new shares that may be purchased by investors. Current shareholders have preferential rights

New capital issue:

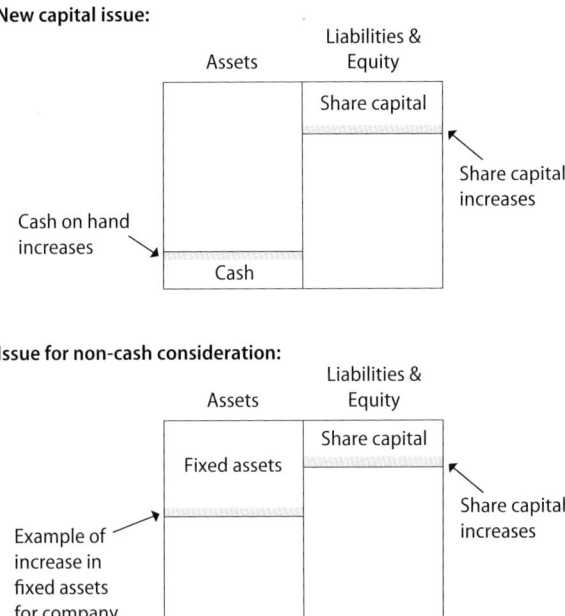

Figure 8.4 New share issues and non-cash share issues.

as far as the purchase of the new shares is concerned (this is often referred to as a *rights issue*). If the share price exceeds the share quota value (i.e., the shares are sold at a *premium*), this difference is accounted for in a share premium account. In practice, most share issues are sold at a premium.

A share issue of, for instance, 1:3 means that a shareholder who has three shares before the new issue is entitled to buy one new share at a fixed price. If the shareholder subscribes to the issue and purchases the new share, he or she then owns four shares. Investors who are not shareholders may purchase subscription rights from shareholders who do not wish to subscribe to the share issue.

Share issues may be directed to specific groups, such as the existing owners, or to specific potential investors, such as pension funds. Such share issues are called *private placements*. If the company is going public by offering the share issue to the general public, this event is generally referred to as an *Initial Public Offering (IPO)*.

In special cases, the new shares are not purchased with cash but rather with some form of property. Such an in-kind share issue is called an *issue for non-cash consideration*. See Figure 8.4. A common example is when a company pays for the purchase of another company with its own shares.

A *stock dividend* (or *share dividend*) is a share issue that does not bring any new financial capital into the company. A stock dividend means that unrestricted equity (for instance, profits brought forward) is converted to restricted equity. See Figure 8.5. A stock dividend only changes the internal structure of the passive side of the balance sheet; there is no change in the total assets or in total equity. A stock dividend of, for instance, 1:2 means that a current shareholder receives one share for every two shares he or she owns (thus, after the stock dividend, the shareholder has three shares for every two shares previously owned). A stock dividend might be issued because the potential financers want to control that the shareholders actually have long-term commitments to the company before they loan the company money. Since it is possible to distribute the unrestricted equity to shareholders, banks and other lenders often require that some portion of the unrestricted equity must be converted to restricted equity in order to lock it into the company.

A company may also split the shares by reducing their quota value. A possible reason for this action is if the market value of the shares is considered too high to be manageable. This procedure is called a *share split*. In a 1:4 split, for instance, current shareholders receive four new shares for one old share. After the split, the shareholder has four shares instead of one share.

Unlisted limited liability companies (the vast majority) cannot raise financial capital by selling shares on the public stock exchanges (e.g., the

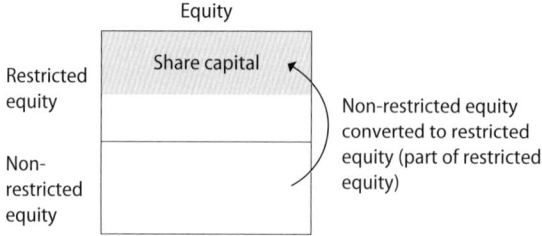

Figure 8.5 Stock dividend issue.

Stockholm Stock Exchange). To finance their operations, they have to turn to various private equity firms, private individuals (referred to as business angels), foundations, or other institutions. In some cases, even public agencies and institutions provide capital.

Borrowing

There are a number of institutions that lend money to companies. Banks, the most important of these institutions, make both short-term and long-term business loans. Typically, banks require companies to offer *collateral* (security) for these loans. Collateral may take different forms. One form is a pledge of a specific property, which means that the lender has the right to that asset and may sell it if the borrower cannot repay the loan. Most banks have relationships with different types of *corporate finance companies.* Such companies can act as lenders in specific situations, for example, when a company needs to extend credit to its customers. Other actors in the credit market are insurance companies.

In consolidated groups, the parent company may lend money to its subsidiaries. These loans are generally *subordinated liabilities.* Even if these loans are not legally subordinated, the parent company, for appearances, may treat them as subordinated.

Large companies may also issue *bonds* as a way to raise financial capital. A large bond issue (bond loan) is divided into a number of equal bonds of, for example, 100 Euros each. The company agrees to pay the bondholders a fixed rate of interest at fixed dates in the year. Typically, bonds are secured by a mortgage on a specific property.

Except for the supplier liabilities (accounts payable), the company's short-term loans are mainly bank loans. One example is a *checking account*, a variable credit that can be used up to a specified maximum limit. Because such checking accounts may continue for several years, they are sometimes listed in the company's long-term liabilities on its balance sheet.

In addition to the collateral terms, other loan terms are the *interest rate,* the *maturity date* of the loan, and the interest and principal payment conditions. These conditions are determined in negotiations between lenders and borrowers. Loan repayments may be structured variously. For example, the loan may either be *amortized* successively over its life (each payment consists of one part principal repayment and one part interest), or the entire loan principal may be due in a lump sum at the maturity date. These terms depend on a number of factors, including the kind of company applying for the loan. A small engineering company probably has little room for negotiating a bank loan with a major bank. On the other hand, when a large, global corporation raises capital, there might be a number of banks competing to offer loans. In this situation, the company has an advantage in the loan negotiations.

Off the balance sheet financing

In the modern era of corporate finance, companies can also finance some of their operations in a way that avoids balance sheet recognition. This is referred to as *off the balance sheet financing*. The most commonly used forms are the following:

- *Factoring* means that the company has borrowed against its invoices (its claims to payment from customers). The invoices are the collateral. In practice, the company exchanges its invoices for cash (at a fee).
- *Leasing* is an alternative to borrowing money, especially for production equipment. The company enters into a lease contract with a financial institution that purchases the equipment and then rents it to the company.
- *Sale/leaseback* is similar to leasing. In this scenario, the company sells an object (e.g., a production plant or a building) to a financial organization and at the same time promises to lease the object for a specific time period. The financial institution has the financial risk (as well as the possibility of increases in the object's value).
- *Outsourcing* is both a way to finance part of a company's operations and a way to manage its operations. Outsourcing basically means that a company pays another company to conduct various aspects of its operations, such as printing, copying, or cleaning. Outsourcing is not a new phenomenon, although it has increased in significance in recent decades. See Chapter 9. The financial objective of outsourcing is to reduce costs as well as capital tied up in fixed assets.

8.4 Cash flow analysis

The *cash flow analysis* allows us to see how a company was financed and how the financial capital flowed through the company during a specific year. The analysis reveals the relationship between company performance (its profit or loss) and the cash with which it paid its invoices, etc. Cash flow analysis also allows us to distinguish between the company's internally and externally generated cash sources and to identify its *working capital needs* (raw materials, products in process, inventories, and accounts receivable), and its *fixed asset cash requirements* (machinery, buildings, etc.). See Section 8.1.

In addition to the balance sheet and the income statement, large and/ or listed limited liability companies are also required to present a *cash flow analysis* in their annual reports. The structure of the cash flow analysis is not formally specified in detail. *Swedish Institute of Authorized Public Accountants (Föreningen Auktoriserade Revisorer,* FAR) has, however, a standardized template for the statement that most Swedish companies use.

The *cash flow analysis* is useful for understanding changes in a company's capital structure. Sometimes another format than the recommended FAR template may be preferred, for instance, for making *financial forecasts.* Whereas financial analysis is made after-the-fact, financial forecasts are made in advance, for example, in order to calculate how much capital to raise, or how many shares to issue. In such cases, the company needs to forecast changes in *working capital* and its *Funds Utilized* and its *Funds Provided* (cf. the presentation in Figure 8.6). Based on this information, the company can make the relevant decisions on borrowing and share issues.

Preparing a cash flow analysis

The first step in preparing a cash flow analysis is to determine the *internally generated funds* from the company's operations. See Figure 8.6. This is calculated from the *profit before appropriations and tax* from the income statement. The year's *depreciation* (a non-cash expense on the income statement) is then added. Looking at the balance sheet, we can see that depreciation reduces total assets through the accumulated depreciation account and reduces the year's profit by the same amount as a cost.

A – FUNDS PROVIDED	
From the year's operations internally generated funds	X
Sale of fixed assets	X
Decrease in long-term receivables	X
New share issues (and other financial contributions)	X
Increase in long-term debts	X
Total funds provided	TOTAL X
B – FUNDS UTILIZED	
Investments in real estate, buildings, machinery, and equipment	Y
Investments in shares and bonds	Y
Increase in long-term receivables	Y
Decrease in long-term debts	Y
Dividends to shareholders	Y
Total funds utilized	TOTAL Y
CHANGE IN WORKING CAPITAL (A – B)	TOTAL X – TOTAL Y
CHANGE IN WORKING CAPITAL, IN DETAIL	
Increase (+) / decrease (–) of inventories	Z
Increase (+) / decrease (–) of short-term receivables	Z
Increase (+) / decrease (–) of liquid assets	Z
Increase (–) / decrease (+) of short-term liabilities	Z
CHANGE IN WORKING CAPITAL	(TOTAL X –TOTAL Y) = TOTAL Z

Figure 8.6 Statement of cash flow presentation per FAR's recommendation.

The second step is to determine the other items under the heading *Funds Provided*. This involves determining the inflow of cash from the sale of fixed assets (e.g., buildings and machinery), from the decrease in long-term receivables (e.g., collections on cash loans to other companies), from new share issues and other financial contributions, and from the increase in long-term debts. These amounts are added to derive the company's Funds Provided.

Then the *Funds Utilized* is determined in a similar way. In this section, we list the cash outflows from purchases of fixed assets (e.g., buildings and machinery), from investments in shares and bonds, from the increase in long-term receivables, from the decrease in long-term debt, and from the dividends to shareholders. These amounts are added to derive the company's Funds Utilized. The difference between the Funds Provided and the Funds Utilized is the change in working capital for the year.

Last, the change in *working capital* is analyzed in detail. This involves adding/subtracting the changes in various working capital accounts (e.g., inventories, short-term receivables, other liquid assets, and short-term liabilities). If the cash flow analysis is prepared correctly, the change in working capital will be the same as the difference between the Funds Provided and the Funds Utilized.

It is important to note that a company's liquidity (i.e., cash on hand) does not necessarily increase with an inflow of new capital (Funds Provided). The amount of the liquid assets also depends on how much of the cash is used (Funds Utilized) and how the other working capital accounts have changed. For example, if a company makes large, long-term investments and increases its inventories, the company's cash on hand may decrease even if it has new cash injections.

8.5 Financial risks

All financial activities involve risks. Investors who buy shares in a company risk losing their investments if the company defaults (e.g., enters bankruptcy). A lender who has made an interest-bearing loan to a company is at less risk if collateral has been pledged as security, but can still lose a substantial amount of the loan. The risks associated with financing a company can be classified in two types:

- operational risk
- financial risk.

A general rule is to avoid investments that combine high operational risk and high financial risk.

Operational risk

Operational risk, or *business risk*, derives from the company's operations (e.g., innovative start up in an emerging industry, or traditional manufacturing of products for a mature market), the character of the operations (e.g., cyclical demand, highly competitive, or politically sensitive), and the company's cost structure and capital tied up. An analysis of operational risk can be made on the basis of the following:

- operations' content and character
- cost structure
- capital tied up.

Key performance index for operational risk

Operational risk can be measured by the return on total capital, ROT, calculated over several years. See Section 7.6. For a company with high operational risk, ROT may vary greatly on an annual basis.

It is important to remember that the return on total capital is a company key performance index. It is not an objective in itself. A company's long-term objective is always that its operations will provide as much return on equity as possible. This objective is applicable to the company as a whole. For different parts of the company operations, this measurement is less relevant. A department head, for example, has very limited ability to influence the return on total capital or on equity, even if his or her actions influence the company performance and its return.

Operational risk: the influence of company activities and character

Let us take the pulp and paper industry as an example of how a company's activities and character affect its operational risk. This industry is sensitive to both domestic and international influence. In the last 20 years, the domestic political pressure to reduce emissions such as sulfur dioxide has clearly influenced certain industrial investments. Moreover, the pulp and paper industry, as a high export industry, is subject to foreign political influence. Fluctuations in the currency exchange rates are also important for machinery and other purchases, and for foreign sales (some of which are marked in US dollars). In short, the operational risk in the pulp and paper industry is significant (although this risk is compensated by its lower financial risk).

Operational risk: the influence of cost structure and tied-up capital

One way to analyze operational risk is to study a company's cost structure and use of capital by applying the *DuPont Chart* (named after the U.S. chemical

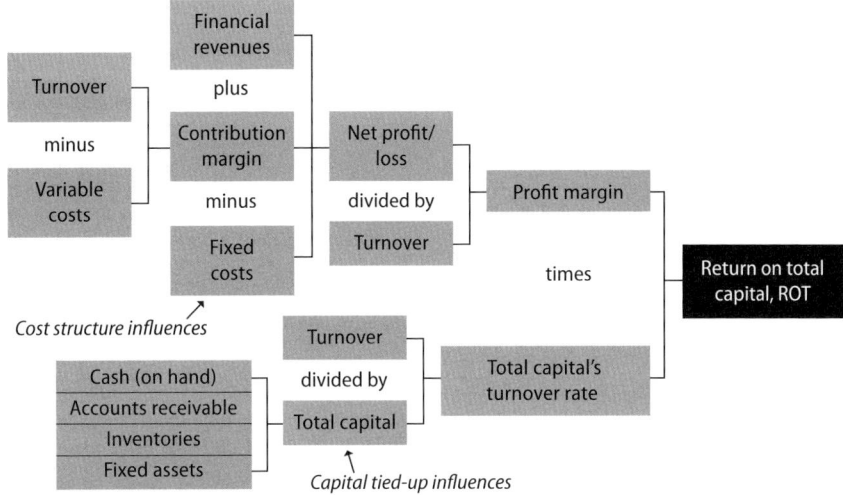

Figure 8.7 The DuPont Chart.

company du Pont de Nemours). Figure 8.7 presents the DuPont Chart in which the return on total capital varies with the company's *cost structure* and *tied-up capital*.

The DuPont Chart reveals how effectively the company's assets are used in their entirety, and how both the tied-up capital and the cost structure influence the return on total capital. Note that no accounts from the liabilities side of the balance sheet are components in the chart. The DuPont Chart describes the company's efficiency independently of its financial funding.

Financial risk

Financial risk focuses on how financial changes affect company operations. An analysis of financial risk can be made on the basis of the following:

- leverage risk
- liquidity risk
- loan duration, interest rate risk, and currency risk.

Leverage risk

Financial risk is often expressed in terms of a number of key ratios. The ratio that is typically used in financial analysis to measure *leverage risk* is the *equity ratio* or the *debt/equity ratio*. See Chapter 7 for the definitions.

The *equity* ratio, which measures the relationship between equity and total capital (i.e., total assets) is a measure of a company's ability to self-finance its operations. The ratio can be used to express how much of the company's total capital is provided by the owners and how much is provided by other actors. A common interpretation of the ratio is that it measures the company's ability to survive in the long term.

In English-speaking countries, instead of the equity ratio, the *debt/equity ratio* is used. This ratio is essentially similar to the equity ratio, although calculated differently.

Leverage equation

$$ROE = ROT + (ROT - R_L) \cdot \frac{D}{E}$$

ROE = return on equity

ROT = return on total capital

R_L = average interest rate on liabilities

$$\frac{D}{E} = \frac{debt}{equity\ ratio}$$

The *leverage equation*, which is used to measure financial risk, shows the relationship between a company's debt/equity ratio, its return (profitability) on total capital and equity, and its average interest rate on its liabilities. A company that has high debt/equity ratio (D/E) according to this formula will have a high return on equity so long as the return on total capital exceeds its average interest rate on its liabilities (in the formula: ROE is large if D/E is large, provided that a ROT > R_L). High D/E can be used as a lever to create a high return for the owners. At the same time, of course, the risk to the return on equity is that high D/E can quickly lead to large negative values if the return on total capital is less than the average interest rate on liabilities (in the formula: ROE turns negatively quickly if D/E is large, provided that ROT < R_L). With lower leverage (a higher equity ratio) the effect is less noticeable, but the risk of its rolling back is reduced also (see the example).

EXAMPLE

Let's take a simple example to analyze the leverage effect. Suppose a company has sales of SEK 110 million, no financial revenues, variable costs of SEK 60 million, fixed costs (excluding interest costs) of SEK 40 million, total assets of SEK 100 million, total equity of SEK 15 million (Total liabilities = Total assets – Equity = SEK 85 million), and an average debt interest rate of 7 %. In other words, the company has a relatively high debt/equity ratio (a low equity ratio).

We begin our analysis using the DuPont Chart, as follows:

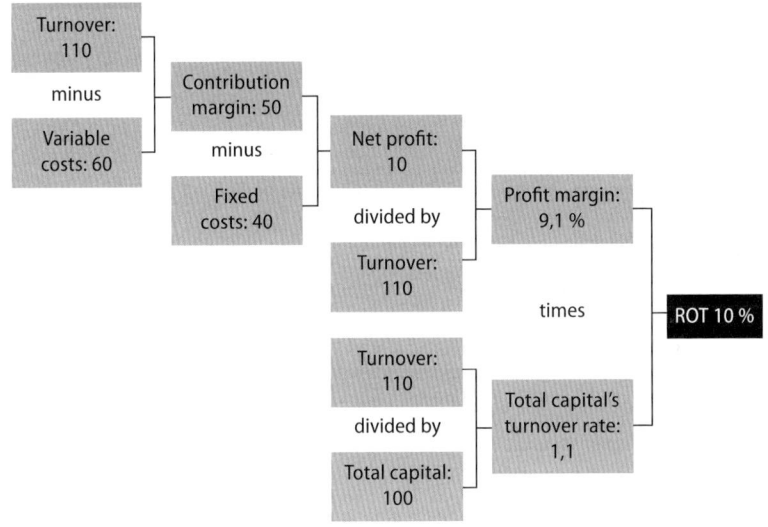

The DuPont Formula calculates:

$$\text{ROE (after tax)} = (100\ \% - \text{tax rate}) \cdot [\text{ROT} + (\text{ROT} - R_L) \cdot \frac{D}{E}] =$$

$$= (100\ \% - 22\ \%) \cdot [10\ \% + (10\ \% - 7\ \%) \cdot \frac{85}{15}] = 21{,}1\ \%$$

Now let us see what a decrease in sales of 10 % would have as far as the leverage effect. We assume that the variable costs follow the decrease in sales, that is, the variable costs also decrease by 10 %. We begin as above with the DuPont Chart:

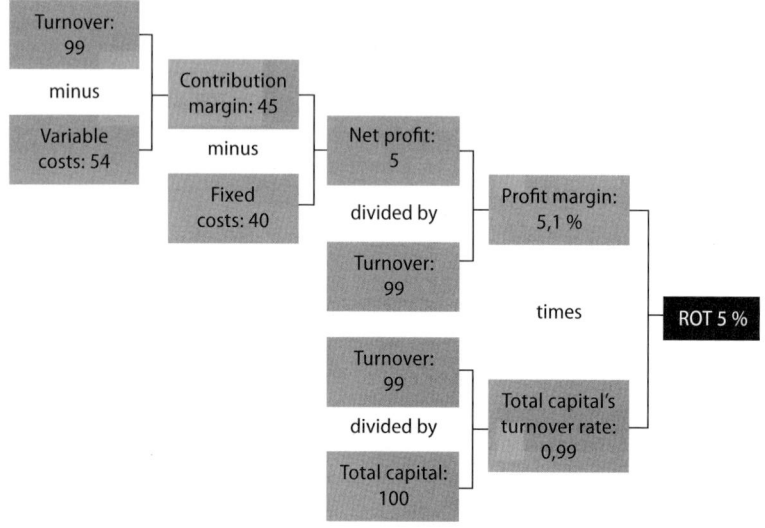

The DuPont Formula calculates:

$$\text{ROE (after tax)} = (100\ \% - 22\ \%) \cdot [5\ \% + (5\ \% - 7\ \%) \cdot \frac{85}{15}] = -4,9\ \%$$

The example shows that even a relatively modest change in the company's revenues can have a powerful leverage effect if the debt/equity ratio is high.

In summary, the leverage equation says that the relationship between liabilities and equity influences the return on equity and on possibility for borrowing money. A large amount of equity makes it more difficult to recoup capital than a smaller amount, but at the same time it poses a lower *credit risk*.

Liquidity risk

Liquidity risk reflects a company's ability to pay its current liabilities in the short term (i.e., the acid test ratio of current assets in relation to current liabilities). See Chapter 7 for the definition.

A company with a poor acid test ratio may have problems in the short term with paying its invoices, salaries, etc. Although the company may be healthy otherwise, and have a high equity ratio, the lack of liquidity is one of the most common causes of loan defaults and bankruptcies. A supplier who does not receive payment for invoices can bring bankruptcy charges against its customer. Such a supplier will be given a higher priority in the repayment list in relation to many other lenders if the company in question actually enters bankruptcy.

It is essential that companies maintain control of their liquidity. For analysis of everyday activities, of course, the liquidity ratios need to be complemented with a number of other tools and techniques. Since a company's cash position can change very quickly, payments must be scheduled, necessary new loans arranged for in advance, and so on. In major companies, cash management is one of the most important responsibilities of the company's Chief Financial Officer (CFO).

Loan duration, interest rate risk, and currency risk

To avoid a shortage of cash (or, on the contrary, accumulating too many non-interest bearing assets), companies need to match the *maturity dates* of their assets and liabilities. By matching we mean that when financing the acquisition of an asset, the liability should come due about the same time that the asset's economic life is expected to end. This is the *debt-maturity principle* that is fundamental in financial risk taking.

Because loan interest rates vary over time, a company that borrows money is exposed to a certain *interest rate risk*. Generally, the company has the option of choosing between a variable rate of interest and a fixed rate of interest. If a significant amount of a company's liabilities (debts) carries a variable rate of interest, the company might see its financial expenses fluctuate considerably over short time periods.

A company is also exposed to *currency risk*. Currency fluctuations can influence both the cash situation and profit. Companies can protect themselves from this risk by buying a *currency hedge,* which is generally a financial instrument (a forward or future contract) that allows the company to buy a specific foreign currency at an agreed-on price at a certain time (or in a certain time period). Of course, currency fluctuations can also be positive as well as negative. For example, if a Swedish exporter sells products abroad, marked in a foreign currency, and the value of the Swedish SEK has fallen relative to that currency when the payment is received, the Swedish exporter's profit will increase (in SEK).

8.6 Two sides of financing

In this chapter, the topic of financing has been discussed primarily as a company's internal activity. However, there is another side to finance. A company's financiers (investors and lenders) take an interest in how the company manages its corporate finances. Financiers often have several investment/lending options, and can choose among companies. They may even choose not to invest in, or lend money to, a company. Besides equities, other options are debt securities (e.g., bonds or certificates of deposit) or commodities (e.g., gold, oil, or pork bellies).

For instance, a company may plan to finance a major investment with long-term loans as well as with a new share issue. Because of the competition from other investment options, the company must present itself as trustworthy and responsible. Therefore, the company will publish an *investment memorandum* (or, in special cases, a more formal *prospectus*) that provides details about the new share issue. In this way, investors receive information on which they can base their decisions. In order to protect the investors, there are legal requirements in most countries, including Sweden, about the presentation and distribution of this kind of information.

Thus, all corporate financing rests on two pillars: control and risk. Investors who want control over a company take a greater investment risk than those who do not. However, those investors who exert control are able to influence company operations and direction. In this position, they will achieve higher returns on their investments if the company is successful.

CHAPTER 9

Industrial operations
– a link in a larger value chain

In this final chapter, we discuss industrial operations in a larger context than we did in Chapter 2 in which the individual company was in focus. Value creation in almost all companies involves one or more value chains or networks in which different companies have different roles in the production and delivery of goods and services. It is important to understand the company's position and role in these value chains. Much value creation occurs outside the company's legal/geographic boundaries, often in some distant part of the world. This situation places very large demands on a company's ability to manage and develop its operations and activities.

9.1 The company's operations in the value chain

An industrial company's operations usually include one or more of the production steps from raw materials to finished goods. Thus, we can identify a number of links in the value creation chain. For example, the production links for a loaf of bread can be traced from planting and harvesting to the flourmill to the wholesaler to the retailer to the baker to the consumer. Although the value really only appears when the consumer eats the bread, value is added at each production link; thus, for example, a loaf of bread has more value than the wheat seeds, than the flour, than the yeast, etc.

Position in the value chain

A company's position in the value chain and the various links in its chain raise strategic questions that influence the company's long-term development and competitiveness. In some industries, individual companies control

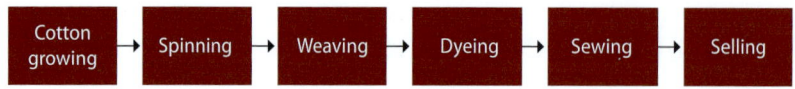

Figure 9.1 The traditional value chain in the clothing industry.

all the value chain links. An example is the oil and gas industry in which large companies control exploration, extraction, refining, and local petrol stations. In other industries, individual companies specialize in only one link in the value chain. An example is a carpentry company that assembles furniture (e.g., IKEA furniture) for other companies.

Figure 9.1 illustrates a single value chain in the clothing industry. The figure shows the links in which an article of clothing (e.g., a pair of jeans) originates with raw materials and moves through the various production steps before reaching the retail store. The clothing industry is an illustrative example of how structural changes in an industry can cause changes in the value chain. Historically, to a great extent the apparel sewing factories controlled the clothing industry. In Sweden, for example, textile companies with brand names such as Algots controlled many links in the chain. Today it is more common that companies between the factories and the retailers control the chain. These companies, which do not manufacture clothes, have developed strong brands. They control large parts of the value chain through marketing, distribution, and design. They are a new type of actor in the textile industry.

One example is H&M. This large, Swedish retailer controls the design, purchase, and sales of its clothes. With its many stores, H&M ensures it has a sales volume large enough to support the relatively low prices of its clothes. By contrast, the Italian clothing retailer, Benetton, does not own the Benetton stores, which are separate legal entities. Despite this structure, Benetton still maintains considerable control over its brand. Benetton has contracts with the stores that require them to sell only Benetton clothes and to follow certain rules on store fittings, etc. In return, the stores benefit from company-wide, joint marketing campaigns. In this way, Benetton enjoys both large and small company advantage.

Another example, also from the clothing industry, is the Spanish company, Zara, which is known for its ownership at all levels: from factories to stores. Zara very effectively collects customer data from stores and operates a highly efficient logistics system that can produce relatively small collections that they can sell at low prices.

IKEA, the Swedish home furnishings corporation, is an example from a different industry. Important links in IKEA's value chain include design, engineering, manufacturing, sales, transport, and assembly. IKEA controls the value chain, but it is not directly involved in all activities. For example, while IKEA has its own product designers, it also uses external designers. The engineering activity is also very important because of the many functional requirements of the IKEA products. IKEA's ready-to-assemble furniture, which is of relatively high quality, must be engineered so that it fits in a (preferably flat) carton. It is important that the flat cartons are as small as possible because they are piled on pallets during transport. Subcontractors in various parts of the world usually manufacture the furnishings and their components. IKEA has stores worldwide – essentially store-warehouse combinations – that sell its products. Typically, customers transport their purchases themselves and then assemble them at home. One reason for IKEA's enormous international success is its creation of a new type of value chain that has changed the entire furniture industry.

Structural changes in the value chain

A value chain may change when a new company enters an industry or when existing companies change their position in the chain. For example, a company may buy another company that is located somewhere along the chain, either "downstream" or "upstream". This is known as *vertical integration*. By contrast, *horizontal integration* occurs when a company buys a competitor positioned in the same link. This type of structural change is relatively common.

Ownership changes, however, are not the only way in which the value chain may change. In the past, large industrial companies often controlled many of the links in the value chain; there was a high degree of vertical integration. In this way, companies could control the quality and delivery of their components by manufacturing them in-house. A well-known example comes from the automotive industry. In the early 20th century, Ford Motor Company owned all the production steps for the Model T – from iron mines to steelworks to metal works to component production to final assembly. The only step not controlled was retail sales by dealers. Several such vertically integrated companies exist today. Zara (see above), for example, has high vertical integration.

However, in recent decades, the trend is more towards *vertical disintegration*. This means that companies are specializing more in their core businesses and leaving earlier steps in the value chain (e.g., manufacturing) to others, such as external suppliers.

Outsourcing

The purchase of activities – that a company has previously performed itself – from another source (e.g., a subcontractor) is called *outsourcing*. Originally, support activities such as printing, transport, company restaurants, computer operations, and facilities management were outsourced. Today, however, companies commonly acquire many more goods and services for their core activities from external sources. In this way, some of the largest and most advanced industrial companies have become systems integrators that orchestrate a wide range of goods and services in their production systems without actually owning them.

Thus, there is a trend today of outsourcing internal operations and not building internal competences and resources for certain steps in the value chain from the beginning. Many large companies employ outside consultants who work at the companies' locations, use the companies' engineering and IT systems, and, to some extent, are thought of as employees. Similarly, people from staffing agencies may work in any number of capacities at companies (although they are employees of the agencies, not the companies). These include people working in production, storage, reception, administration, etc. The boundaries between a company's activities and operations and its legal borders may thus differ markedly.

Insourcing

At the same time as many companies specialize in some activities and outsource other activities, there still is a strong desire to take control of the value chain by controlling what were traditionally viewed as customer functions.

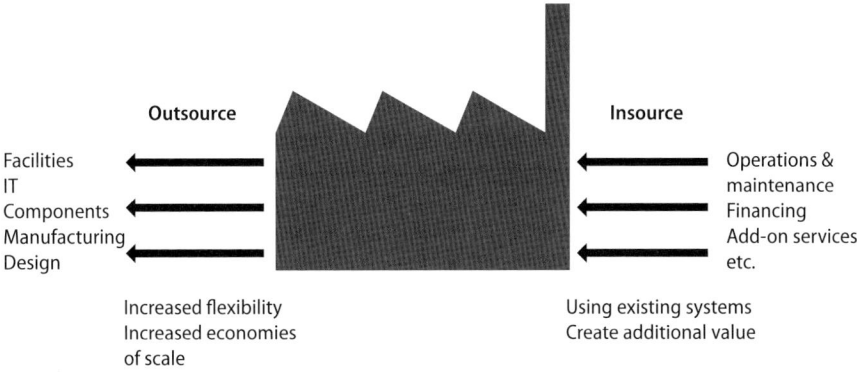

Outsource

Facilities
IT
Components
Manufacturing
Design

Insource

Operations &
maintenance
Financing
Add-on services
etc.

Increased flexibility
Increased economies
of scale

Using existing systems
Create additional value

Figure 9.2 The current trend forward in the industrial value chains.

This is called *insourcing*. Thus, many companies have employees at customer locations who do maintenance and service work and install equipment and systems. Companies often develop and manage operations of systems for their customers – systems that they originally delivered. See Figure 9.2.

One example is the Swedish company, Ericsson, which not only develops telecommunication equipment and builds telecommunication systems but also manages systems operations for various telecom operators. Borrowing a word from the computer industry, people sometimes talk about a company's *installed base* (i.e., the equipment and systems that the company has delivered and are used by various operators). This installed base creates new possibilities for the development and sale of other items, for example, spare parts, other equipment and supplies, and, most importantly, various services that add to the value creation for customers (and increase revenues for the supplier; see Chapter 2). In this way, a company binds its customers ever closer.

Thus, constant changes, movements, and trends influence the value chain. Different strategic trends have been popular at different times. One reason that many companies try to position themselves further down the value chain is that *value added* in operations with convergent production flows (see Chapter 2) is typically greater in the later links than in the earlier links. The value added in the earlier links is usually low. A sheep farmer in

Scotland, for example, receives only a very small part of the price of a wool sweater sold in Sweden.

Although value-added is usually greatest in the later production stages (downstream in the value chain), the same may not be true for a company's profitability. Very often the competition is fierce, and the marketing costs are high at the end of the value chain (e.g., as in the clothing industry). However, other industry sectors may differ. For example, in the oil and gas industry the greatest value-added occurs in the early production stages, upstream in the value chain. As a result, financial investors and professional trading companies now own many petrol stations that the oil and gas companies previously owned. Today, the range of products and services at large petrol stations is quite different from previous years.

Value chains and networks

To this point, we have discussed rather simple and individual examples of *value chains*. In practice, value chains are much more complex, in part because various value chains link to each other. Moreover, a single company may be part of several value chains. If a company's finished products consist of many different components, the picture is very complicated indeed. The European steel producer, Ovako, sells special steel to SKF, the world's leader in the manufacture and sale of ball bearings. From this steel, SKF manufactures ball bearings for car manufacturers (e.g., Volvo) and companies in other sectors. Ovako also sells steel directly to Volvo's subcontractors and to other companies. Both SKF and Ovako are participants in a great many value chains. Therefore, our definition of a value chain must be broadened to consider these situations.

A *network* is another way to describe the interaction between companies. The network concept emphasizes the relationships between different companies. In many situations, one can draw a very extensive network of companies that are, in some way, integrated. The difference between a value chain and a network derives mainly from their different starting points. The network emphasizes actors, activities, and resources. The value chain emphasizes the transformation in the various links in the value chain.

9.2 The company's supply chain

Supply chain management

As the result of specialization in the value chain, at many companies the purchasing department's function has become increasingly important. At the same time, the production manager has become responsible for sourcing, not only for internal production but also for external outsourcing and sub-contractor production. From the perspective of the individual company, one talks of the *supply chain*, and the management of the company's purchasing from suppliers as *supply chain management.*

Figure 9.3 is a simplified picture of the supply chain in which the company's value creation is linked to suppliers and customers. Raw materials, components, information, ideas, and sometimes people flow along the supply chain's network of supplier-customer relationships. At the supplier end of the chain, a company typically has a number of suppliers of raw materials and other goods and services.

For many companies, *contract manufacturers* use the company's draw-ings, designs, and specifications to manufacture their products (see H&M and IKEA above). This phenomenon is often also found in other sectors such as the electronics industry.

It is also common that relatively complex components are developed and manufactured by external actors, some of which are large, global subcontrac-tors with their own brands and products. Two examples of such companies are SKF (see above) and Bosch, the German manufacturer of, among other things, automotive components. Although these large companies sell com-ponents to many customers, these sales usually are only a small percentage of their customers' total product.

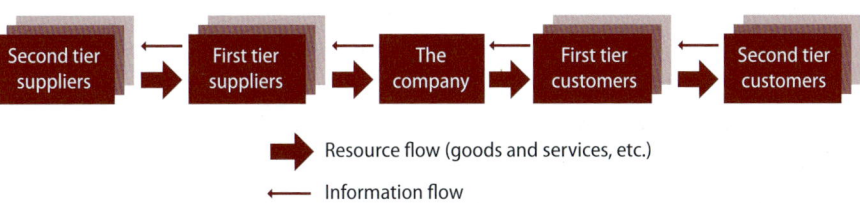

Figure 9.3 The company's supply chain.

Many subcontractors also have their own suppliers for their operations. Thus, the supply chain may be very long, with first tier suppliers, second tier suppliers, and so on. The same situation exists for customers who have their own customers, who have their own customers, and so on. The tighter these supply chains are, the more efficient they are. At the same time, however, problems can spread along the supply chain. For example, if a subcontractor does not supply a component, the entire downstream supply chain slows or stops. In the same way, if customer demands and buying habits change, the upstream supply chain for production might have a problem. Examples of important decisions in supply chain management are the following: supply chain design, sourcing and management of suppliers, and in-house production vs. supplier production.

Offshoring, re-shoring, and next shoring

Issues around the location of subcontractors and suppliers and their activities are closely associated with a company's supply chain management. In recent decades, a strong trend has emerged of locating both manufacturing and engineering activities in low cost countries. Advanced developments in web-based communication, telecommunications, and logistics have led to increased globalization of industrial activities. Geography is no longer the barrier it once was. For example, now it is possible to send technical information immediately to partners on the other side of the globe.

IT systems and telecommunication technology have, in many ways, made companies' operations location-independent. Service activities such as call centers and telecommunication-based service centers may be located thousands of miles from the customers who use them. Similarly, many labor-intensive production operations have moved from the old, rich industrialized countries to countries in Asia, Africa, and Eastern Europe. The same is true of software design and development. Asian and Eastern European countries now have at least as many qualified engineers as Western European countries and the United States. Movement of operations such as this is called *offshoring* (not to be confused with outsourcing, explained above).

The advantages and disadvantages of offshoring are much debated. While the short-term economic benefits of offshoring are often obvious, critics point out that a company's ability to innovate and develop usually triumphs in the long term because of the geographic proximity to product development, production, and marketing.

In recent years, a counter-trend has arisen in which companies return operations that were previously off-shored to their home countries. There are several reasons for *re-shoring*. For example, the benefits of the low cost countries may have disappeared as those countries become wealthier. Second, a company may find the complexity of managing operations and maintaining high quality production in distant regions or countries too difficult. Third, technological developments may have advanced to the point where the costs of machinery and equipment have sunk low enough to restart production in a company's home country.

In *next shoring,* a company moves its manufacturing to hubs or centers of demand and innovation. In this model, companies locate to where local or regional demand is high, and a supply of innovative suppliers is present. Therefore, in the future we may expect that companies will continue to choose production locations on the basis of efficiency and economic considerations, wherever these locations are.

Value creation as a boundary-spanning activity

Concepts such as *supply chain, supply chain management, outsourcing,* and *offshoring* together reveal an important feature in modern industrialization: value creation often spans a company's judicial boundaries. A company's partners often handle important activities in the development, production, and distribution of its goods and services. Customers generally are largely unaware of, and probably indifferent to, the long list of companies behind the goods and services they purchase. However, it is crucial to understand a company's supply chains and its networks that involve all these external actors if we are to understand how value is created.

REFERENCES

Cusumano, M.A., Kahl, S.J., & Suarez, F.F. (2015), "Services, Industry Evolution, and the Competitive Strategies of Product Firms", *Strategic Management Journal*, 36(4), 559–575.

Fayol, H. (1916*), Industrial and General Administration* (English translation 1930). London: Sir Isaac Pitman & Sons.

Galbraith, J. (1973), *Designing Complex Organizations*. Reading: Addison-Wesley.

Hayes, R.H., & Wheelwright, S.C. (1979), "Link Manufacturing Process and Product Life Cycles", *Harvard Business Review*, 57(1), 133–140.

Johnston, R., & Clark, G. (2008), *Service Operations Management: Improving Service Delivery* (3 ed.). Harlow: FT/Prentice Hall.

Lantz, B., Isaksson, A., & Löfsten, H. (2014), *Industriell ekonomi: Grundläggande ekonomisk analys*. Lund: Studentlitteratur.

Mintzberg, H. (1979), *The Structuring of Organizations*. Englewood Cliffs: Prentice Hall.

Appendix

DICTIONARY FOR TERMS IN INDUSTRIAL MANAGEMENT: ENGLISH–SWEDISH

A-share A-aktie
absorption costing påläggskalkylering
accounting plan kontoplan
accounts payable leverantörsskuld
accounts receivable kundfordran
accrued costs upplupna kostnader
accrued revenues upplupna intäkter
accumulated additional
    depreciations ackumulerade
    överavskrivningar
acid test ratio kassalikviditet
activity based costing ABC-kalkylering
additional depreciations överavskrivning
additional depreciations avskrivning
    över plan
additional sales tilläggsförsäljning
adjusted equity justerat eget kapital
administrative overhead (AO) administ-
    rationsomkostnad (AO)
advance payment förskott
aftermarket eftermarknad
allocation to a particular time
    period periodisering
amortization amortering
Annual accounts
    act Årsredovisningslagen
annual accounts book årsbok
annual financial statements årsbokslut
annual meeting bolagsstämma
annual profit/loss årets vinst/förlust

annual report årsredovisning
annuity annuitet
annuity factor annuitetsfaktor
annuity method annuitetsmetoden
annuity ratio annuitetskvot
appropriations bokslutsdispositioner
articles of incorporation bolagsordning
asset tillgång
asset account tillgångskonto
asset side aktivsidan
auditor revisor
auditor's report revisionsberättelse
authority befogenhet
authorized public accountant auktori-
    serad revisor
average interest on debt genomsnittlig
    skuldränta

B-share B-aktie
balance account balanskonto
balance carried forward utgående balans
balance sheet balansräkning
bank balances banktillgodohavande
basis for overhead charge påläggsbas
board of directors styrelse
bond obligation
bond loan obligationslån
Book keeping act Bokföringslagen
break-even chart resultatdiagram
business administration företagsekonomi

business concept affärsidé
business model affärsmodell
business ratio nyckeltal
business transaction affärshändelse

capital gain realisationsvinst
capital investment
  analysis investeringsbedömning
capital investment
  appraisal investeringskalkyl
capital requirements kapitalbehov
capital stock aktiekapital
cash (in hand) kassa, likvida medel
cash flow analysis finansieringsanalys,
  kassaflödesanalys
cash flow forecast finansieringsprognos
cash payment (out) utbetalning
cash payment surplus
  utbetalningsöverskott
cash receipt (in) inbetalning
cash receipt
  surplus inbetalningsöverskott
change of working capital förändring av
  rörelsekapital
checking account checkkredit
chief executive officer (CEO) verkställande direktör (VD)
civil law civillagstiftning
closing balance utgående balans
code kontera
common cost samkostnad
common stock stamaktier
companies act Aktiebolagslagen
conditions for liability (debt)
  payment återbetalningsvillkor
consolidated balance
  sheet koncernbalansräkning
consolidated financial
  statement koncernredovisning
consolidated income
  statement koncernresultaträkning
consumables förbrukningsvaror

consumer products konsumentprodukter
contract manufacturer legotillverkare
contribution costing bidragskalkyl
contribution margin täckningsbidrag
convertible instrument konvertibler
convertible loan konvertibla förlagslån
coordination samordning
core business kärnverksamhet
corporate income tax bolagsskatt
cost kostnad
cost account kostnadskonto
cost center kostnadsställe
cost driver kostnadsdrivare
cost of acquisition anskaffningsvärde
cost of capital (cost for
  financing) kapitalkostnad
cost of capital (interest rate) kalkylränta
cost of goods sold kostnad sålda varor
cost of production tillverkningskostnad
cost unit kalkylobjekt, kostnadsbärare
credit kredit
credit rating kreditvärdering
credit risk kreditrisk
creditor fordringsägare
critical volume kritisk volym
cumulative present value nusumma
cumulative present value
  factor nusummefaktor
current assets omsättningstillgångar
current liabilities kortfristiga skulder
current ratio balanslikviditet
current recording of transactions löpande
  bokföring

debit debet
debt/equity ratio skuldsättningsgrad
debts skulder
deferred income förutbetalda intäkter
deferred tax liabilities latent skatteskuld
depreciation avskrivningar
depreciation according to plan planenlig
  avskrivning, avskrivningar enligt plan

depreciation as recorded in the books räkenskapsenliga avskrivningar
direct cost direkt kostnad
direct labor cost direkt lön
direct material cost direkt material
dividend utdelning
division of labor arbetsfördelning
double taxation dubbelbeskattning
double-entry bookkeeping system dubbel bokföring
DuPont chart DuPont-schema

earnings vinst
Economic Association Act Lagen om ekonomiska föreningar
economic cycle ekonomiskt kretslopp
economic life ekonomisk livslängd
economics nationalekonomi
economy of scale stordriftsfördelar
effectiveness yttre effektivitet
efficiency inre effektivitet
employers social cost for labor arbetsgivaravgift
equity eget kapital
equity method kapitalandelsmetoden (i koncernredovisning)
equity method reserve kapitalandelsfond
equity ratio soliditet
equity/assets ratio soliditet
equity/debt ratio soliditet
estimate förkalkyl
ex post costing efterkalkyl
expenditure utgift
expenditure account utgiftskonto
expense kostnad
externally generated funds externt genererade medel
extra ordinary costs (expenses) extraordinära kostnader
extra ordinary income extraordinära intäkter

finance company finansbolag
financial accounting affärsredovisning, extern redovisning
financial analysis räkenskapsanalys
financial cost finansiell kostnad
financial ratio finansiella nyckeltal
financial revenue finansiell intäkt
first tier supplier förstaledsleverantör
fixed asset cash requirement anläggningskapitalbehov
fixed assets anläggningstillgångar
fixed cost fast kostnad
formal structure formell organisationsstruktur
full costing method självkostnadsmetoden
functional staff structure linje–stabsorganisation
functional manager linjechef
functional structure funktionsorganisation, linjeorganisation
functional work supervision funktionell arbetsledning
funds provided tillförda medel
funds utilized använda medel

gearing skuldsättningsgrad
general ledger huvudbok
generally accepted accounting principles god redovisningssed
geographical structure geografisk organisation
goods varor
goodwill goodwill
group (consolidated company) koncern
group contribution koncernbidrag

income inkomst
income statement resultaträkning
income statement classified by function funktionsindelad resultaträkning

**income statement classified by nature
of cost** kostnadsslagsindelad
resultaträkning
**index for distributing
overhead** fördelningsnyckel
indirect cost indirekt kostnad
industrial management industriell
ekonomi
industry industri, bransch
informal structure informell organisation
input market anskaffningsmarknad
input VAT ingående moms
installed base installerad bas av sålda
produkter
interest ränta
interest rate risk ränterisk
interest-bearing liability räntebärande
skulder
interim claim interimsfordran
interim debt interimsskuld
internal financing självfinansiering
internal rate of return internränta
**internal rate of return
method** internräntemetoden
internally generated funds internt
genererade medel
internally provisioned funds internt
tillförda medel
inventory färdigvarulager, förråd
investment investering
investor finansiär
invoice faktura
**issue for non-cash consideration
(in kind)** apportemission

job description befattningsbeskrivning
journal dagbok

**knowledge intensive business
service** kunskapsintensiv
tjänsteverksamhet

labor intensive services personalintensiva
tjänster
land assets markanläggningar
leverage equation hävstångsformeln
leverage risk hävstångsrisk
liabilities skulder
liability account skuldkonto
**liability side (in the financial
statement)** passivsidan
(i balansräkningen)
line organization linjeorganisation
loan interest låneränta
long-term debt långfristig skuld
lower of cost or market lägsta värdets
princip

machinery and equipment maskiner och
inventarier
management accounting
internredovisning
management report
förvaltningsberättelse
managing director verkställande
direktör
**manufacturing
overhead** tillverkningsomkostnad
manufacturing overhead charge tillverkningsomkostnadspålägg
marketing marknadsföring
marketing process
marknadsföringsprocessen
material overhead materialomkostnad
material overhead charge
materialomkostnadspålägg
matrix structure matrisorganisation
mortgage pantförskrivning
municipal tax law (in Sweden)
Kommunalskattelagen

net present ratio nuvärdekvot
net present value nuvärde

net present value method nuvärdemetoden, kapitalvärdemetoden
net sales nettoförsäljning,
nettoomsättning
new capital issue nyemission
nominal value nominellt värde
non-restricted equity fritt eget kapital

obsolescence inkurans
opening balance ingående balans
operating cost rörelsens kostnader
operating volume verksamhetsvolym
operational investment
rationaliseringsinvestering
operational liabilities rörelseskulder
operational risk rörelserisk
operations verksamhet, produktion,
drift
organization chart organisationsplan
organization design organisationsform
organization structure
organisationsstruktur
original investment grundinvestering
other direct manufacturing cost övriga
direkta tillverkningskostnader
other operational costs övriga
rörelsekostnader
other operational revenues övriga
rörelseintäkter
output market avsättningsmarknader
output VAT utgående moms
overhead charge omkostnadspålägg
overhead cost omkostnad
owners capital, equity riskbärande eget
kapital
owners capital, equity riskkapital

parent company moderbolag
payback method återbetalningsmetoden,
pay-back metoden
payback period återbetalningstid

payment utbetalning
performance measurement prestationsmätning, resultatmätning
pooling of interest method poolningsmetoden (inom koncernredovisning)
preferred stock preferensaktier
prepaid expenses förutbetalda kostnader
present value nuvärde
present value factor nuvärdefaktor
private placement riktad emission
process costing divisionskalkyl
product cost estimate produktkalkyl
product costing produktkalkylering
product development produktutveckling
product structure produktorganisation
product variety produktvariation
production flow produktionsflödet
productivity produktivitet
products in progress produkter i arbete
profit vinst
profit (or loss) brought forward
balanserade vinst-(förlust-)medel
profit / loss resultat
profit and loss account vinst- och
förlustkonto
profit center resultatenhet
profit charge vinstpålägg
profit margin vinstmarginal
profit/loss account resultatkonto
profitability lönsamhet
project structure (organization)
projektorganisation
proportional method klyvningsmetoden
(inom koncernredovisning)
purchase accounting method förvärvsmetoden (inom koncernredovisning)

quarterly report kvartalsrapport

raw materials inventory råmaterialförråd
real estate fastighet

real estate tax fastighetsskatt
receipt inbetalning
receivable fordran
replacement cost återanskaffningsvärde
required return avkastningskrav
residual value restvärde
restricted equity bundet eget kapital
restricted shares bundna aktier
return räntabilitet, avkastning
return on capital employed räntabilitet på sysselsatt kapital
return on equity räntabilitet på eget kapital
return on total capital räntabilitet på totalt kapital
revaluation reserve uppskrivningsfond
revenue intäkt
revenue account intäktskonto
revenue unit intäktsenhet
rights issue företrädesemission

safety margin säkerhetsmarginal
sales omsättning, försäljning
sales & administrative overhead försälj-nings- och administrationskostnad
sales overhead försäljningsomkostnad
sales revenue försäljningsintäkter
second tier supplier andraledsleverantör
service business tjänsteverksamheter
services tjänster
share aktier
share capital aktiekapital
share issue aktieemission
share premium reserve överkursfond
share register aktiebok
shareholder aktieägare
small business fåmansföretag
social security costs sociala avgifter
span of control kontrollspann
special direct sales costs speciella direkta försäljningskostnader

specific cost särkostnad
specific revenue särintäkt
staffing agency bemanningsföretag
stakeholder intressent
standing committee stående kommitté
step costing stegkalkyl
stock dividend fondemission
stock issue aktieemission
strategic investment inriktningsinvestering
structure of equity and liabilities kapitalstruktur
subordinated loan efterställt lån, subordinerat lån
subsidiary dotterbolag
supplier underleverantör
supply chain försörjningskedja
Swedish Accounting Standards Board Bokföringsnämnden
Swedish Company Registration Office Bolagsverket
Swedish Financial Reporting Board Rådet för finansiell rapportering
Swedish Financial Supervisory Authority Finansinspektionen
Swedish Institute of Authorized Public Accountants Föreningen auktori-serade revisorer (FAR)
Swedish Tax Agency Skatteverket

T-account T-konto
task force tillfällig arbetsgrupp
tax skatt
tax allocation reserve periodiseringsfond
tax debt skatteskuld
tax law skattelagstiftning
tax return deklaration
taxpayer skattesubjekt
technical life teknisk livslängd
terminal value slutvärde
terminal value factor slutvärdefaktor

total assets totalt kapital,
balansomslutning
transfer pricing internprissättning
turnover omsättning, försäljning
turnover rate omsättningshastighet
type of cost kostnadsslag

unappropriated equity disponibelt eget
kapital
unit sales enhetsförsäljning
unrestricted shares fria aktier
untaxed reserves obeskattade reserver

valuation of assets värdering av
tillgångar
valuation of current assets värdering av
omsättningstillgångar
valuation of fixed assets värdering av
anläggningstillgångar
valuation of liabilities (debts) värdering
av skulder

valuation of time tidspreferens
value added förädlingsvärde
value added tax (VAT) mervärdesskatt
(moms)
value chain värdekedja
value creation värdeskapande
value depreciation värdeminskning
value proposition värdeerbjudande
variable cost rörlig kostnad
venture capital company riskkapitalbolag
vertical integration vertikal integration
voucher verifikation

working capital rörelsekapital
working capital
change rörelsekapitalförändring
working capital
requirement rörelsekapitalbehov

year end closing bokslut
yield avkastning

DICTIONARY FOR TERMS IN INDUSTRIAL MANAGEMENT: SWEDISH–ENGLISH

A-aktie A-share
ABC-kalkylering activity based costing
ackumulerade överavskrivningar accumulated additional depreciations
administrationsomkostnad (AO) administrative overhead (AO)
affärshändelse business transaction
affärsidé business concept
affärsmodell business model
affärsredovisning financial accounting
aktiebok share register
Aktiebolagslagen companies act
aktieemission stock issue, share issue
aktiekapital share capital, capital stock
aktier share
aktieägare shareholder
aktivsidan asset side
amortering amortization
andraledsleverantörer second tier supplier
anläggningskapitalbehov fixed asset cash requirement
anläggningstillgångar fixed assets
annuitet annuity
annuitetsfaktor annuity factor
annuitetskvot annuity ratio
annuitetsmetoden annuity method
anskaffningsmarknad input market
anskaffningsvärde cost of acquisition

använda medel funds utilized
apportemission issue for non-cash consideration (in kind)
arbetsfördelning division of labor
arbetsgivaravgift employers social cost for labor
auktoriserad revisor authorized public accountant
avkastning return, yield
avkastningskrav required return
avskrivning över plan additional depreciations
avskrivningar depreciation
avskrivningar enligt plan depreciation according to plan
avsättningsmarknader output market

B-aktie B-share
balanserade vinst-(förlust-)medel profit (or loss) brought forward
balanskonto balance account
balanslikviditet current ratio
balansomslutning total assets
balansräkning balance sheet
banktillgodohavande bank balances
befattningsbeskrivning job description
befogenhet authority
bemanningsföretag staffing agency
bidragskalkyl contribution costing

Bokföringslagen Book keeping act
Bokföringsnämnden Swedish
 Accounting Standards Board
bokslut year end closing
bokslutsdispositioner appropriations
bolagsordning articles of incorporation
bolagsskatt corporate income tax
bolagsstämma annual meeting
Bolagsverket Swedish Company
 Registration Office
bransch industry
bundet eget kapital restricted equity
bundna aktier restricted shares

checkkredit checking account
civillagstiftning civil law

dagbok journal
debet debit
deklaration tax return
direkt kostnad direct cost
direkt lön direct labor cost
direkt material direct material cost
disponibelt eget kapital unappropriated
 equity
divisionskalkyl process costing
dotterbolag subsidiary
drift operations
dubbel bokföring double-entry
 bookkeeping system
dubbelbeskattning double taxation
DuPont-schema DuPont chart

efterkalkyl ex post costing
eftermarknad aftermarket
efterställt lån subordinated loan
eget kapital equity
ekonomisk livslängd economic life
ekonomiskt kretslopp economic cycle
enhetsförsäljning unit sales
extern redovisning financial accounting

externt genererade medel externally
 generated funds
extraordinära intäkter extra ordinary
 income
extraordinära kostnader extra ordinary
 cost, (expenses)

faktura invoice
fast kostnad fixed cost
fastighet real estate
fastighetsskatt real estate tax
finansbolag finance company
finansiell intäkt financial revenue
finansiell kostnad financial cost
finansiella nyckeltal financial ratio
finansieringsanalys cash flow analysis
finansieringsprognos cash flow forecast
Finansinspektionen Swedish Financial
 Supervisory Authority
finansiär investor
fondemission stock dividend
fordran receivable
fordringsägare creditor
formell organisationsstruktur formal
 structure
fria aktier unrestricted shares
fritt eget kapital non-restricted equity
funktionell arbetsledning functional
 work supervision
funktionsindelad resultaträkning income
 statement classified by function
funktionsorganisation functional
 structure
fåmansföretag small business
färdigvarulager inventories
förbrukningsvaror consumables
fördelningsnyckel index for distributing
 overhead
Föreningen auktoriserade revisorer
 (FAR) Swedish Institute of
 Authorized Public Accountants

företagsekonomi business administration
företrädesemission rights issue
förkalkyl estimate
förråd inventories
förskott advance payment
förstaledsleverantör first tier supplier
försäljning sales, turnover
försäljnings- och administrations- kostnad sales & administrative overhead
försäljningsintäkter sales revenue
försäljningsomkostnad sales overhead
försörjningskedja supply chain
förutbetalda intäkter deferred income
förutbetalda kostnader prepaid expenses
förvaltningsberättelse management report
förvärvsmetoden (inom koncernredo- visning) purchase accounting method
förädlingsvärde value added
förändring av rörelsekapital change of working capital

genomsnittlig skuldränta average interest on debt
geografisk organisation geographical structure
god redovisningssed generally accepted accounting principles
goodwill goodwill
grundinvestering original investment

huvudbok general ledger
hävstångsformeln leverage equation
hävstångsrisk leverage risk

inbetalning receipt, cash receipt (in)
inbetalningsöverskott cash receipt surplus
indirekt kostnad indirect cost
industri industry

industriell ekonomi industrial management
informell organisation informal structure
ingående balans opening balance
ingående moms input VAT
inkomst income
inkurans obsolescence
inre effektivitet efficiency
inriktningsinvestering strategic investment
installerad bas av sålda produkter installed base
interimsfordran interim claim
interimsskuld interim debt
internprissättning transfer pricing
internredovisning management accounting
internränta internal rate of return
internräntemetoden internal rate of return method
internt genererade medel internally generated funds
internt tillförda medel internally provisioned funds
intressent stakeholder
intäkt revenue
intäktsenhet revenue unit
intäktskonto revenue account
investering investment
investeringsbedömning capital investment analysis
investeringskalkyl capital investment appraisal

justerat eget kapital adjusted equity

kalkylobjekt cost unit
kalkylränta cost of capital (interest rate)
kapitalandelsfond equity method reserve

kapitalandelsmetoden (inom koncernredo-visning) equity method
kapitalbehov capital requirements
kapitalkostnad cost of capital (cost for financing)
kapitalstruktur structure of equity and liabilities
kapitalvärdemetoden net present value method
kassa cash (in hand)
kassaflödesanalys cash flow analysis
kassalikviditet acid test ratio
klyvningsmetoden (inom koncernredo-visning) proportional method
Kommunalskattelagen municipal tax law (in Sweden)
koncern group (consolidated company)
koncernbalansräkning consolidated balance sheet
koncernbidrag group contribution
koncernredovisning consolidated financial statement
koncernresultaträkning consolidated income statement
konsumentprodukter consumer products
kontera code
kontoplan accounting plan
kontrollspann span of control
konvertibla förlagslån convertible loan
konvertibler convertible instrument
kortfristiga skulder current liabilities
kostnad cost
kostnad expense
kostnad sålda varor cost of goods sold
kostnadsbärare cost unit
kostnadsdrivare cost driver
kostnadskonto cost account
kostnadsslag type of cost
kostnadsslagsindelad resultat-räkning income statement classified by nature of cost
kostnadsställe cost center

kredit credit
kreditrisk credit risk
kreditvärdering credit rating
kritisk volym critical volume
kundfordran accounts receivable
kunskapsintensiv tjänsteverksamhet knowledge intensive business service
kvartalsrapport quarterly report
kärnverksamhet core business

Lagen om ekonomiska föreningar Economic Association Act
latent skatteskuld deferred tax liabilities
legotillverkare contract manufacturer
leverantörsskuld accounts payable
likvida medel cash (in hand)
linje–stabsorganisation functional staff structure
linjechef functional manager
linjeorganisation line organization, functional structure
låneränta loan interest
långfristig skuld long-term debt
lägsta värdets princip lower of cost or market
lönsamhet profitability
löpande bokföring current recording of transactions

markanläggningar land assets
marknadsföring marketing
marknadsföringsprocessen marketing process
maskiner och inventarier machinery and equipment
materialomkostnad material overhead
materialomkostnadspålägg material overhead charge
matrisorganisation matrix structure
mervärdesskatt (moms) value added tax (VAT)
moderbolag parent company

nationalekonomi economics
nettoförsäljning net sales
nettoomsättning net sales
nominellt värde nominal value
nusumma cumulative present value
nusummefaktor cumulative present
 value factor
nuvärde present value, net present value
nuvärdefaktor present value factor
nuvärdekvot net present ratio
nuvärdemetoden net present value
 method
nyckeltal business ratio
nyemission new capital issue

obeskattade reserver untaxed reserves
obligation bond
obligationslån bond loan
omkostnad overhead cost
omkostnadspålägg overhead charge
omsättning sales, turnover
omsättningshastighet turnover rate
omsättningstillgångar current assets
organisationsform organization design
organisationsplan organization chart
organisationsstruktur organization
 structure

pantförskrivning mortgage
passivsidan (i balansräkningen) liability
 side (in the financial statement)
pay-back metoden payback method
periodisering allocation to a particular
 time period
periodiseringsfond tax allocation reserve
personalintensiva tjänster labor intensive
 services
**poolningsmetoden (inom koncernredo-
visning** pooling of interest method
planenlig avskrivning depreciation
 according to plan
preferensaktier preferred stock

prestationsmätning performance
 measurement
produkter i arbete products in progress
produktion operations
produktionsflödet production flow
produktivitet productivity
produktkalkyl product cost estimate
produktkalkylering product costing
produktorganisation product structure
produktutveckling product development
produktvariation product variety
projektorganisation project structure
 (organization)
pålåggsbas basis for overhead charge
pålåggskalkylering absorption costing

rationaliseringsinvestering operational
 investment
realisationsvinst capital gain
restvärde residual value
resultat profit/loss
resultatdiagram break-even chart
resultatenhet profit center
resultatkonto profit/loss account
resultatmätning performance
 measurement
resultaträkning income statement
revisionsberättelse auditor's report
revisor auditor
riktad emission private placement
riskbärande eget kapital owners capital,
 equity
riskkapital owners capital, equity
riskkapitalbolag venture capital
 company
Rådet för finansiell rapportering Swedish
 Financial Reporting Board
råmaterialförråd raw materials inventory
räkenskapsanalys financial analysis
räkenskapsenliga avskrivningar depreciation as recorded in the books
ränta interest

räntabilitet return
räntabilitet på eget kapital return on equity
räntabilitet på sysselsatt kapital return on capital employed
räntabilitet på totalt kapital return on total capital
räntebärande skulder interest-bearing liability
ränterisk interest rate risk
rörelsekapital working capital
rörelsekapitalbehov working capital requirement
rörelsekapitalförändring working capital change
rörelsens kostnader operating cost
rörelserisk operational risk
rörelseskulder operational liabilities
rörlig kostnad variable cost

samkostnad common cost
samordning coordination
självfinansiering internal financing
självkostnadsmetoden full costing method
skatt tax
skattelagstiftning tax law
skatteskuld tax debt
skattesubjekt taxpayer
Skatteverket Swedish Tax Agency
skulder debts, liabilities
skuldkonto liability account
skuldsättningsgrad debt/equity ratio, gearing
slutvärde terminal value
slutvärdefaktor terminal value factor
sociala avgifter social security costs
soliditet equity ratio, equity/assets ratio
speciella direkta försäljnings-kostnader special direct sales costs
stamaktier common stock
stegkalkyl step costing

stordriftsfördelar economy of scale
styrelse board of directors
stående kommitté standing committee
subordinerat lån subordinated loan
säkerhetsmarginal safety margin
särintäkt specific revenue
särkostnad specific cost

T-konto T-account
teknisk livslängd technical life
tidspreferens valuation of time
tillfällig arbetsgrupp task force
tillförda medel funds provided
tillgång asset
tillgångskonto asset account
tillverkningskostnad cost of production
tillverkningsomkostnad manufacturing overhead
tillverkningsomkostnadspålägg manufacturing overhead charge
tilläggsförsäljning additional sales
tjänster services
tjänsteverksamheter service business
totalt kapital total assets
tvingande investering required investment
täckningsbidrag contribution margin

underleverantör supplier
upplupna intäkter accrued revenues
upplupna kostnader accrued costs
uppskrivningsfond revaluation reserve
utbetalning payment, cash payment (out)
utbetalningsöverskott cash payment surplus
utdelning dividend
utgift expenditure
utgiftskonto expenditure account
utgående balans balance carried forward, closing balance

utgående balans closing balance
utgående moms output VAT

varor goods
verifikation voucher
verksamhet operations
verksamhetsvolym operating volume
verkställande direktör (VD) chief
executive officer (CEO), managing
director
vertikal integration vertical integration
vinst earnings, profit
vinst- och förlustkonto profit and loss
account
vinstmarginal profit margin
vinstpålägg profit charge
värdeerbjudande value proposition
värdekedja value chain
värdeminskning value depreciation
värdering av anläggningstillgångar
valuation of fixed assets
värdering av omsättningstillgångar
valuation of current assets
värdering av skulder valuation
of liabilities (debts)

värdering av tillgångar valuation of assets
värdeskapande value creation

yttre effektivitet effectiveness

årets vinst/förlust annual profit/loss
årsbok annual accounts book
årsbokslut annual financial statements
årsredovisning annual report
Årsredovisningslagen Annual accounts
act
återanskaffningsvärde replacement cost
återbetalningsmetoden payback method
återbetalningstid payback period
återbetalningsvillkor conditions for
liability (debt) payment

överavskrivning additional depreciations
överkursfond share premium reserve
övriga direkta tillverkningskostnader
other direct manufacturing cost
övriga rörelseintäkter other operational
revenues
övriga rörelsekostnader other
operational costs

INTEREST TABLES

Table A Future value (FV) $(1+i)^y$

Year	0%	1%	2%	3%	4%	5%	6%	7%	8%	9%	10%	11%	12%	13%	14%	15%	16%	17%	18%	20%
1	1,0	1,010	1,020	1,030	1,040	1,050	1,060	1,070	1,080	1,090	1,100	1,110	1,120	1,130	1,140	1,150	1,160	1,170	1,180	1,200
2	1,0	1,020	1,040	1,061	1,082	1,103	1,124	1,145	1,166	1,188	1,210	1,232	1,254	1,277	1,300	1,323	1,346	1,369	1,392	1,440
3	1,0	1,030	1,061	1,093	1,125	1,158	1,191	1,225	1,260	1,295	1,331	1,368	1,405	1,443	1,482	1,521	1,561	1,602	1,643	1,728
4	1,0	1,041	1,082	1,126	1,170	1,216	1,262	1,311	1,360	1,412	1,464	1,518	1,574	1,630	1,689	1,749	1,811	1,874	1,939	2,074
5	1,0	1,051	1,104	1,159	1,217	1,276	1,338	1,403	1,469	1,539	1,611	1,685	1,762	1,842	1,925	2,011	2,100	2,192	2,288	2,488
6	1,0	1,062	1,126	1,194	1,265	1,340	1,419	1,501	1,587	1,677	1,772	1,870	1,974	2,082	2,195	2,313	2,436	2,565	2,700	2,986
7	1,0	1,072	1,149	1,230	1,316	1,407	1,504	1,606	1,714	1,828	1,949	2,076	2,211	2,353	2,502	2,660	2,826	3,001	3,185	3,583
8	1,0	1,083	1,172	1,267	1,369	1,477	1,594	1,718	1,851	1,993	2,144	2,305	2,476	2,658	2,853	3,059	3,278	3,511	3,759	4,300
9	1,0	1,094	1,195	1,305	1,423	1,551	1,689	1,838	1,999	2,172	2,358	2,558	2,773	3,004	3,252	3,518	3,803	4,108	4,435	5,160
10	1,0	1,105	1,219	1,344	1,480	1,629	1,791	1,967	2,159	2,367	2,594	2,839	3,106	3,395	3,707	4,046	4,411	4,807	5,234	6,192
11	1,0	1,116	1,243	1,384	1,539	1,710	1,898	2,105	2,332	2,580	2,853	3,152	3,479	3,836	4,226	4,652	5,117	5,624	6,176	7,430
12	1,0	1,127	1,268	1,426	1,601	1,796	2,012	2,252	2,518	2,813	3,138	3,498	3,896	4,335	4,818	5,350	5,936	6,580	7,288	8,916
13	1,0	1,138	1,294	1,469	1,665	1,886	2,133	2,410	2,720	3,066	3,452	3,883	4,363	4,898	5,492	6,153	6,886	7,699	8,599	10,699
14	1,0	1,149	1,319	1,513	1,732	1,980	2,261	2,579	2,937	3,342	3,797	4,310	4,887	5,535	6,261	7,076	7,988	9,007	10,147	12,839
15	1,0	1,161	1,346	1,558	1,801	2,079	2,397	2,759	3,172	3,642	4,177	4,785	5,474	6,254	7,138	8,137	9,266	10,539	11,974	15,407
16	1,0	1,173	1,373	1,605	1,873	2,183	2,540	2,952	3,426	3,970	4,595	5,311	6,130	7,067	8,137	9,358	10,748	12,330	14,129	18,488
17	1,0	1,184	1,400	1,653	1,948	2,292	2,693	3,159	3,700	4,328	5,054	5,895	6,866	7,986	9,276	10,761	12,468	14,426	16,672	22,186
18	1,0	1,196	1,428	1,702	2,026	2,407	2,854	3,380	3,996	4,717	5,560	6,544	7,690	9,024	10,575	12,375	14,463	16,879	19,673	26,623
19	1,0	1,208	1,457	1,754	2,107	2,527	3,026	3,617	4,316	5,142	6,116	7,263	8,613	10,197	12,056	14,232	16,777	19,748	23,214	31,948
20	1,0	1,220	1,486	1,806	2,191	2,653	3,207	3,870	4,661	5,604	6,727	8,062	9,646	11,523	13,743	16,367	19,461	23,106	27,393	38,338

Table B Present Value (PV) $1/((1+i)^y)$

Year	0 %	1 %	2 %	3 %	4 %	5 %	6 %	7 %	8 %	9 %	10 %	11 %	12 %	13 %	14 %	15 %	16 %	17 %	18 %	20 %
1	1,000	0,990	0,980	0,971	0,962	0,952	0,943	0,935	0,926	0,917	0,909	0,901	0,893	0,885	0,877	0,870	0,862	0,855	0,847	0,833
2	1,000	0,980	0,961	0,943	0,925	0,907	0,890	0,873	0,857	0,842	0,826	0,812	0,797	0,783	0,769	0,756	0,743	0,731	0,718	0,694
3	1,000	0,971	0,942	0,915	0,889	0,864	0,840	0,816	0,794	0,772	0,751	0,731	0,712	0,693	0,675	0,658	0,641	0,624	0,609	0,579
4	1,000	0,961	0,924	0,888	0,855	0,823	0,792	0,763	0,735	0,708	0,683	0,659	0,636	0,613	0,592	0,572	0,552	0,534	0,516	0,482
5	1,000	0,951	0,906	0,863	0,822	0,784	0,747	0,713	0,681	0,650	0,621	0,593	0,567	0,543	0,519	0,497	0,476	0,456	0,437	0,402
6	1,000	0,942	0,888	0,837	0,790	0,746	0,705	0,666	0,630	0,596	0,564	0,535	0,507	0,480	0,456	0,432	0,410	0,390	0,370	0,335
7	1,000	0,933	0,871	0,813	0,760	0,711	0,665	0,623	0,583	0,547	0,513	0,482	0,452	0,425	0,400	0,376	0,354	0,333	0,314	0,279
8	1,000	0,923	0,853	0,789	0,731	0,677	0,627	0,582	0,540	0,502	0,467	0,434	0,404	0,376	0,351	0,327	0,305	0,285	0,266	0,233
9	1,000	0,914	0,837	0,766	0,703	0,645	0,592	0,544	0,500	0,460	0,424	0,391	0,361	0,333	0,308	0,284	0,263	0,243	0,225	0,194
10	1,000	0,905	0,820	0,744	0,676	0,614	0,558	0,508	0,463	0,422	0,386	0,352	0,322	0,295	0,270	0,247	0,227	0,208	0,191	0,162
11	1,000	0,896	0,804	0,722	0,650	0,585	0,527	0,475	0,429	0,388	0,350	0,317	0,287	0,261	0,237	0,215	0,195	0,178	0,162	0,135
12	1,000	0,887	0,788	0,701	0,625	0,557	0,497	0,444	0,397	0,356	0,319	0,286	0,257	0,231	0,208	0,187	0,168	0,152	0,137	0,112
13	1,000	0,879	0,773	0,681	0,601	0,530	0,469	0,415	0,368	0,326	0,290	0,258	0,229	0,204	0,182	0,163	0,145	0,130	0,116	0,093
14	1,000	0,870	0,758	0,661	0,577	0,505	0,442	0,388	0,340	0,299	0,263	0,232	0,205	0,181	0,160	0,141	0,125	0,111	0,099	0,078
15	1,000	0,861	0,743	0,642	0,555	0,481	0,417	0,362	0,315	0,275	0,239	0,209	0,183	0,160	0,140	0,123	0,108	0,095	0,084	0,065
16	1,000	0,853	0,728	0,623	0,534	0,458	0,394	0,339	0,292	0,252	0,218	0,188	0,163	0,141	0,123	0,107	0,093	0,081	0,071	0,054
17	1,000	0,844	0,714	0,605	0,513	0,436	0,371	0,317	0,270	0,231	0,198	0,170	0,146	0,125	0,108	0,093	0,080	0,069	0,060	0,045
18	1,000	0,836	0,700	0,587	0,494	0,416	0,350	0,296	0,250	0,212	0,180	0,153	0,130	0,111	0,095	0,081	0,069	0,059	0,051	0,038
19	1,000	0,828	0,686	0,570	0,475	0,396	0,331	0,277	0,232	0,194	0,164	0,138	0,116	0,098	0,083	0,070	0,060	0,051	0,043	0,031
20	1,000	0,820	0,673	0,554	0,456	0,377	0,312	0,258	0,215	0,178	0,149	0,124	0,104	0,087	0,073	0,061	0,051	0,043	0,037	0,026

Table C Cumulative Present Value (CPV) $(1-(1+i)^{-n})/i$

Year	0%	1%	2%	3%	4%	5%	6%	7%	8%	9%	10%	11%	12%	13%	14%	15%	16%	17%	18%	20%
1		0,990	0,980	0,971	0,962	0,952	0,943	0,935	0,926	0,917	0,909	0,901	0,893	0,885	0,877	0,870	0,862	0,855	0,847	0,833
2		1,970	1,942	1,913	1,886	1,859	1,833	1,808	1,783	1,759	1,736	1,713	1,690	1,668	1,647	1,626	1,605	1,585	1,566	1,528
3		2,941	2,884	2,829	2,775	2,723	2,673	2,624	2,577	2,531	2,487	2,444	2,402	2,361	2,322	2,283	2,246	2,210	2,174	2,106
4		3,902	3,808	3,717	3,630	3,546	3,465	3,387	3,312	3,240	3,170	3,102	3,037	2,974	2,914	2,855	2,798	2,743	2,690	2,589
5		4,853	4,713	4,580	4,452	4,329	4,212	4,100	3,993	3,890	3,791	3,696	3,605	3,517	3,433	3,352	3,274	3,199	3,127	2,991
6		5,795	5,601	5,417	5,242	5,076	4,917	4,767	4,623	4,486	4,355	4,231	4,111	3,998	3,889	3,784	3,685	3,589	3,498	3,326
7		6,728	6,472	6,230	6,002	5,786	5,582	5,389	5,206	5,033	4,868	4,712	4,564	4,423	4,288	4,160	4,039	3,922	3,812	3,605
8		7,652	7,325	7,020	6,733	6,463	6,210	5,971	5,747	5,535	5,335	5,146	4,968	4,799	4,639	4,487	4,344	4,207	4,078	3,837
9		8,566	8,162	7,786	7,435	7,108	6,802	6,515	6,247	5,995	5,759	5,537	5,328	5,132	4,946	4,772	4,607	4,451	4,303	4,031
10		9,471	8,983	8,530	8,111	7,722	7,360	7,024	6,710	6,418	6,145	5,889	5,650	5,426	5,216	5,019	4,833	4,659	4,494	4,192
11		10,368	9,787	9,253	8,760	8,306	7,887	7,499	7,139	6,805	6,495	6,207	5,938	5,687	5,453	5,234	5,029	4,836	4,656	4,327
12		11,255	10,575	9,954	9,385	8,863	8,384	7,943	7,536	7,161	6,814	6,492	6,194	5,918	5,660	5,421	5,197	4,988	4,793	4,439
13		12,134	11,348	10,635	9,986	9,394	8,853	8,358	7,904	7,487	7,103	6,750	6,424	6,122	5,842	5,583	5,342	5,118	4,910	4,533
14		13,004	12,106	11,296	10,563	9,899	9,295	8,745	8,244	7,786	7,367	6,982	6,628	6,302	6,002	5,724	5,468	5,229	5,008	4,611
15		13,865	12,849	11,938	11,118	10,380	9,712	9,108	8,559	8,061	7,606	7,191	6,811	6,462	6,142	5,847	5,575	5,324	5,092	4,675
16		14,718	13,578	12,561	11,652	10,838	10,106	9,447	8,851	8,313	7,824	7,379	6,974	6,604	6,265	5,954	5,668	5,405	5,162	4,730
17		15,562	14,292	13,166	12,166	11,274	10,477	9,763	9,122	8,544	8,022	7,549	7,120	6,729	6,373	6,047	5,749	5,475	5,222	4,775
18		16,398	14,992	13,754	12,659	11,690	10,828	10,059	9,372	8,756	8,201	7,702	7,250	6,840	6,467	6,128	5,818	5,534	5,273	4,812
19		17,226	15,678	14,324	13,134	12,085	11,158	10,336	9,604	8,950	8,365	7,839	7,366	6,938	6,550	6,198	5,877	5,584	5,316	4,843
20		18,046	16,351	14,877	13,590	12,462	11,470	10,594	9,818	9,129	8,514	7,963	7,469	7,025	6,623	6,259	5,929	5,628	5,353	4,870
25		22,023	19,523	17,413	15,622	14,094	12,783	11,654	10,675	9,823	9,077	8,422	7,843	7,330	6,873	6,464	6,097	5,766	5,467	4,948
30		25,808	22,396	19,600	17,292	15,372	13,765	12,409	11,258	10,274	9,427	8,694	8,055	7,496	7,003	6,566	6,177	5,829	5,517	4,979
35		29,409	24,999	21,487	18,665	16,374	14,498	12,948	11,655	10,567	9,644	8,855	8,176	7,586	7,070	6,617	6,215	5,858	5,539	4,992
40		32,835	27,355	23,115	19,793	17,159	15,046	13,332	11,925	10,757	9,779	8,951	8,244	7,634	7,105	6,642	6,233	5,871	5,548	4,997
45		36,095	29,490	24,519	20,720	17,774	15,456	13,606	12,108	10,881	9,863	9,008	8,283	7,661	7,123	6,654	6,242	5,877	5,552	4,999
50		39,196	31,424	25,730	21,482	18,256	15,762	13,801	12,233	10,962	9,915	9,042	8,304	7,675	7,133	6,661	6,246	5,880	5,554	4,999

Table D Annuity (ANN) $i/(1-(1+i)^{-y})$

Year	0%	1%	2%	3%	4%	5%	6%	7%	8%	9%	10%	11%	12%	13%	14%	15%	16%	17%	18%	20%
1	1,00	1,010	1,020	1,030	1,040	1,050	1,060	1,070	1,080	1,090	1,100	1,110	1,120	1,130	1,140	1,150	1,160	1,170	1,180	1,200
2	0,50	0,508	0,515	0,523	0,530	0,538	0,545	0,553	0,561	0,568	0,576	0,584	0,592	0,599	0,607	0,615	0,623	0,631	0,639	0,655
3	0,33	0,340	0,347	0,354	0,360	0,367	0,374	0,381	0,388	0,395	0,402	0,409	0,416	0,424	0,431	0,438	0,445	0,453	0,460	0,475
4	0,25	0,256	0,263	0,269	0,275	0,282	0,289	0,295	0,302	0,309	0,315	0,322	0,329	0,336	0,343	0,350	0,357	0,365	0,372	0,386
5	0,20	0,206	0,212	0,218	0,225	0,231	0,237	0,244	0,250	0,257	0,264	0,271	0,277	0,284	0,291	0,298	0,305	0,313	0,320	0,334
6	0,17	0,173	0,179	0,185	0,191	0,197	0,203	0,210	0,216	0,223	0,230	0,236	0,243	0,250	0,257	0,264	0,271	0,279	0,286	0,301
7	0,14	0,149	0,155	0,161	0,167	0,173	0,179	0,186	0,192	0,199	0,205	0,212	0,219	0,226	0,233	0,240	0,248	0,255	0,262	0,277
8	0,13	0,131	0,137	0,142	0,149	0,155	0,161	0,167	0,174	0,181	0,187	0,194	0,201	0,208	0,216	0,223	0,230	0,238	0,245	0,261
9	0,11	0,117	0,123	0,128	0,134	0,141	0,147	0,153	0,160	0,167	0,174	0,181	0,188	0,195	0,202	0,210	0,217	0,225	0,232	0,248
10	0,10	0,106	0,111	0,117	0,123	0,130	0,136	0,142	0,149	0,156	0,163	0,170	0,177	0,184	0,192	0,199	0,207	0,215	0,223	0,239
11	0,09	0,096	0,102	0,108	0,114	0,120	0,127	0,133	0,140	0,147	0,154	0,161	0,168	0,176	0,183	0,191	0,199	0,207	0,215	0,231
12	0,08	0,089	0,095	0,100	0,107	0,113	0,119	0,126	0,133	0,140	0,147	0,154	0,161	0,169	0,177	0,184	0,192	0,200	0,209	0,225
13	0,08	0,082	0,088	0,094	0,100	0,106	0,113	0,120	0,127	0,134	0,141	0,148	0,156	0,163	0,171	0,179	0,187	0,195	0,204	0,221
14	0,07	0,077	0,083	0,089	0,095	0,101	0,108	0,114	0,121	0,128	0,136	0,143	0,151	0,159	0,167	0,175	0,183	0,191	0,200	0,217
15	0,07	0,072	0,078	0,084	0,090	0,096	0,103	0,110	0,117	0,124	0,131	0,139	0,147	0,155	0,163	0,171	0,179	0,188	0,196	0,214
16	0,06	0,068	0,074	0,080	0,086	0,092	0,099	0,106	0,113	0,120	0,128	0,136	0,143	0,151	0,160	0,168	0,176	0,185	0,194	0,211
17	0,06	0,064	0,070	0,076	0,082	0,089	0,095	0,102	0,110	0,117	0,125	0,132	0,140	0,149	0,157	0,165	0,174	0,183	0,191	0,209
18	0,06	0,061	0,067	0,073	0,079	0,086	0,092	0,099	0,107	0,114	0,122	0,130	0,138	0,146	0,155	0,163	0,172	0,181	0,190	0,208
19	0,05	0,058	0,064	0,070	0,076	0,083	0,090	0,097	0,104	0,112	0,120	0,128	0,136	0,144	0,153	0,161	0,170	0,179	0,188	0,206
20	0,05	0,055	0,061	0,067	0,074	0,080	0,087	0,094	0,102	0,110	0,117	0,126	0,134	0,142	0,151	0,160	0,169	0,178	0,187	0,205

INDEX

interim claim 162
interim report 160
internally generated funds 201
internal rate of return 114
internal rate of return method 114
inventory turnover rate 191
investment 97
 – adaptation 120
 – calculation 102
 – calculation methods 118
 – mandatory 121
 – memorandum 210
 – non-financial consequences 123
 – non-mandatory 121
 – original 99
 – rationalization 120
 – strategic 120
invoice 126

journal 138

Kreuger crash 180

leasing 200
leverage equation 206
liability 131
 – current 152
 – discharge from 160
 – interest-bearing 194
 – interim 162
 – non-current 152
 – operational 195
 – side 151
 – subordinated 194
 – tax 152
liaison role 63
licensing 41
limited liability company 148
liquidity 176
logic
 – back office-dominant 32

 – front office-dominant 32
 – goods-dominant 32

management 14
management accounting 142
Managing Director (MD) 49, 149
marketing 38
matrix structure 64
modularized product 37
Municipal Tax Act 146
mutual adjustment 48

net present ratio 109
net present value method 103
net sales 155
net turnover 155
network 66, 216
next shoring 219

offshoring 218
off the balance sheet financing 200
operations 38
 – industrial 23
 – service 18
 – technology-intensive 19
organic growth 196
organization 45
 – chart 49
 – conflict 68
 – control 52
 – formal 45, 50
 – informal 50
 – line 59
 – line-staff 60
 – norms 47
 – objectives 68
 – project 64
 – roles 47
 – structure 46, 49
 – values 47
original investment (OI) 99